Dolomite Reservoirs

Geochemical Techniques for Evaluating Origin and Distribution

by

J. R. Allan
Allan Geological Analysis
Fullerton, California, U.S.A.

and

W. D. Wiggins
Chevron Petroleum Technology Company
La Habra, California, U.S.A.

AAPG Continuing Education Course Note Series #36

Published by
The American Association of Petroleum Geologists
Tulsa, Oklahoma, U.S.A.
Printed in the U.S.A.

Published November, 1993 by the
Education Department of
The American Association of Petroleum Geologists
Tulsa, Oklahoma, U.S.A. 74101

ISBN: 0-89181-185-0

This book, and other titles in the Education Course Note Series, are available from:

The AAPG Bookstore
P.O. Box 979
Tulsa, OK 74101-0979

Telephone: 1-800-364-AAPG (U.S. only)
(918) 584-2555
FAX: (918) 584-0469

PREFACE

Dolomite Reservoirs was written to address the need for a short, clear text that explains commonly used inorganic geochemical techniques and their application to dolomite petroleum reservoirs. It was written in order to transfer this useful geochemical technology to exploration and development geologists in a form that facilitates easy application of geochemical concepts.

Dolomitization results in unique reservoir geometries that bear directly on exploration strategy and field development. An understanding of the process that controlled dolomitization is helpful for predicting reservoir location, geometry, and continuity. Inorganic chemistry can often provide that understanding. Because no text is available to lead the petroleum geologist through the different geochemical techniques, earth scientists must go directly to the research literature to seek information. Unfortunately, the literature is full of studies in which geochemistry has been used improperly. This volume covers many state-of-the-art approaches to the study of dolomitization, and will help geoscientists to critically evaluate geochemical data and interpretations and to design their own geochemical studies of dolomite units.

Part I consists of chapters on different geochemical techniques, with guidelines on how best to apply them, interpret the data, and recognize and avoid the pitfalls and misconceptions that one commonly encounters. Many of the techniques described in Part I can be applied with equal success to limestone reservoir diagenesis.

Part II consists of case studies of dolomite petroleum reservoirs that formed in each of the major dolomitization environments. The examples were selected to illustrate the ways in which geochemistry can be used in conjunction with stratigraphic, structural, petrographic, and facies data to identify the origin of dolomite reservoirs. These case studies should serve as useful templates for the earth scientist who wishes to use geochemical data to evaluate dolomite reservoirs and to determine which dolomite model best fits his or her reservoir or play.

We hope that this book will help geoscientists understand better the many ways in which geochemistry can be used to address dolomite reservoir problems, and will encourage geologists to apply geochemical techniques to petroleum exploration and exploitation in carbonate provinces.

J. R. ALLAN
W. D. WIGGINS

Publisher's Note

❦

AAPG extends its appreciation to the authors, and to Chevron Petroleum Technology Company for releasing this material for publication.

TABLE OF CONTENTS

INTRODUCTION 1
DOLOMITE MODELS 2
DOLOMITE PETROGRAPHY 4

PART I. INORGANIC GEOCHEMICAL TECHNIQUES
ANALYTICAL APPROACHES 6
FORMATION WATER ANALYSES 9
STABLE ISOTOPES 9
SOLUTES 12
OXYGEN ISOTOPES IN DOLOMITE 19
APPROACH #1. COMPARISON TO MARINE CARBONATES 21
APPROACH #2. COMPARISON TO OTHER DOLOMITES 27
CALCULATING ISOTOPE TEMPERATURE 31
CARBON ISOTOPES IN DOLOMITE 33
FLUID INCLUSIONS 39
HYDROCARBON IDENTIFICATION 40
HOMOGENIZATION TEMPERATURE (Th) 41
SALINITY (Tm) 43
Th-Tm PLOTS 45
STRONTIUM ISOTOPES 49
THEORY 49
APPLICATIONS 51
TRACE ELEMENTS 53
THEORY 53
CATHODOLUMINESCENCE AND STAINING TECHNIQUES 55

PART II. CASE STUDIES
DOLOMITE ASSOCIATED WITH SUBTIDAL
TO SUPRATIDAL CYCLES 59
RECOGNITION 59
RED RIVER FORMATION 62
SAN ANDRES FORMATION 66
REFLUX DOLOMITE 72
RECOGNITION 73
SMACKOVER FORMATION 73
NIAGARA FORMATION 77
MARINE AND MIXING ZONE DOLOMITE 82
RECOGNITION 84
STE. GENEVIEVE FORMATION 84

TABLE OF CONTENTS (cont.)

BURIAL DOLOMITE 90
RECOGNITION 95
TRENTON FORMATION 96
SAUDI ARABIAN CRETACEOUS AND JURASSIC FORMATIONS 99
NISKU FORMATION 105

REFERENCES 110

APPENDIX I. CONVENTIONS FOR REPORTING ISOTOPE DATA I-1

APPENDIX II. INTERPRETATION OF DOLOMITES BY COMPARISON TO MARINE CARBONATES II-1
COMPARISON TO MARINE INVERTEBRATES II-1
COMPARISON TO MARINE CARBONATE CEMENTS II-1
LIMITATIONS OF THE TECHNIQUE II-2
CONSTRUCTION OF $\delta^{18}O$ BASELINES II-3
CONSTRUCTION OF $\delta^{13}C$ BASELINES II-3

APPENDIX III. INTERPRETATION OF DOLOMITES BY COMPARISON TO OTHER DOLOMITES III-1
DESCRIPTION OF THE TECHNIQUE III-1
LIMITATIONS OF THE TECHNIQUE III-3

INTRODUCTION

INTRODUCTION

In carbonate systems, dolomite often forms the best reservoirs. The dolomitization of carbonate sediments and rocks increases crystal size and pore throat size, and decreases pore roughness. Without exception, the combination of these increases the permeability of the carbonate. In this way, dolomitization "makes" the reservoir, with dolomite serving as the reservoir and the surrounding limestone forming the seal. Porosity in the dolomitic facies is enhanced in such cases not so much as a result of dolomitization, but through leaching of skeletal grains or evaporites, or by fracturing.

Many environments of dolomitization have been identified. Some result in unique reservoir geometries that bear directly on exploration strategy. Therefore, it is important to gain a firm understanding of the process which controlled dolomitization early in the exploration history of a basin.

In addition, dolomite is less reactive than calcite, so dolomite units are more resistant to porosity loss with depth than limestone units (Fig. 1) and the depth to "economic basement" is commonly greater for dolomite than for limestone. Therefore, the spatial distribution of dolomitized intervals within

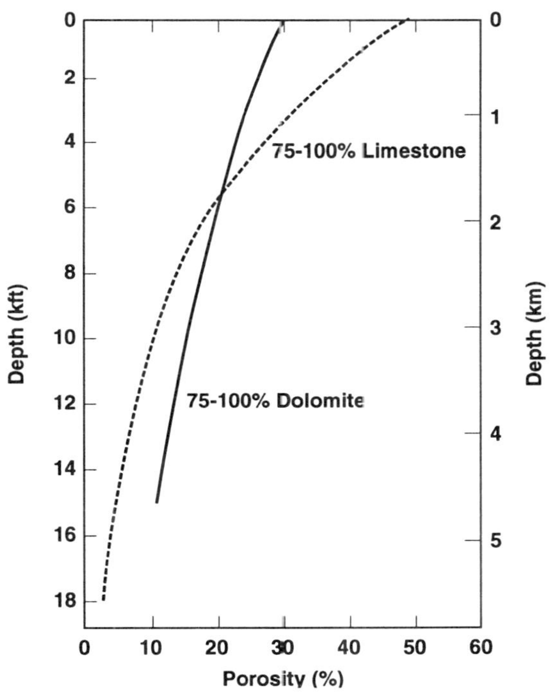

Modified after Schmoker and Halley, 1982

Figure 1. Plot of average porosity vs depth for limestones and dolomites of the South Florida basin. Porosity data are derived from borehole-gravity measurements and from acoustic, neutron, and density logs. Although the range of porosity values can be considerable at any given depth, the average porosity vs depth trends for limestone and dolomite are quite different. Above 5600 ft, limestones are, on the average, more porous than dolomites. Below 5600 ft, dolomites are, on the average, more porous than limestones, because they are less susceptible to porosity reduction by diagenesis and recrystallization than limestones.

a carbonate section often defines the limits of reservoir development. The hydrologic process that dolomitizes a limestone can control the morphology of the dolomite body as well (e.g., supratidal dolomitization can produce thin, stratified, laterally extensive reservoirs, whereas subsurface fault-controlled dolomitization can produce narrow, linear, vertically extensive reservoirs). Therefore, to predict the spatial distribution of a dolomite reservoir, it is advisable to first determine the process that produced the dolomite.

Our goal in inorganic geochemistry is to improve our understanding of the processes which formed a dolomite reservoir in a way which will aid reservoir delineation. For example, geochemical evidence for hydrothermal dolomite may increase the importance of recognizing deep subsurface faults, which can act as conduits for dolomitizing fluids and can be identified using seismic reflection profiles and well data. Evidence for low-temperature, hypersaline dolomite may focus attention on mapping sabkha depositional facies from core and identifying the positions of ancient strandlines. Geochemical techniques succeed when they improve our understanding of the origin of a dolomite reservoir and therefore favor a particular exploration model. Although geochemistry sometimes provides equivocal results, in our experience, failure most often occurs because of lack of integration of multiple types of evidence and an over-reliance on a single technique (e. g., stable isotopes). Inorganic geochemistry is most valuable when it constrains dolomitization models and helps to focus attention on probable reservoir targets. It should always be used in conjunction with other methods of investigation, which, when integrated, may provide information on where a geologist should spot a well.

This manual can be used as a guide by the geologist planning to start a geochemical study of dolomitization, or as an aid to interpreting geochemical data already collected. Because of the length of this manual and the complexity of the topics covered, it is *not* intended to be read from beginning to end as a treatise. Instead, it has been set up as a series of stand-alone chapters. Part I consists of chapters on each of the geochemical techniques commonly applied to dolomite problems. Part II consists of case studies of dolomite reservoirs that formed in each of the major dolomitization environments, and illustrates how each environment can be recognized using the techniques described in Part I in conjunction with other forms of data.

DOLOMITE MODELS

In order to dolomitize a limestone, two requirements must be met: (1) a source of Mg sufficient to dolomitize the limestone must exist; and (2) a transport mechanism must be available to carry the Mg to the site of dolomitization, because mass balance calculations indicate that hundreds of pore volumes of fluid must be exchanged in order to complete the dolomitization reaction. In most cases, the source of Mg is seawater or a seawater-derived formation water or brine. Thus, identifying the hydrologic system responsible for "pumping" the dolomitizing fluid through a limestone is the most important step in identifying the origin of a dolomite. Before commencing with our discussion of geochemical techniques, we will briefly introduce each of the basic dolomitization models. More complete discussions of these models can be found in Morrow (1982b), Land (1986), and Morse and Mackenzie (1990). Those most important to petroleum reservoir formation are described in greater detail in Part II of the manual.

Dolomite can form in or near hypersaline environments from seawater with a Mg/Ca ratio which has been greatly elevated by evaporite precipitation. Precipitation of gypsum and anhydrite, in particular, can preferentially remove Ca from seawater while leaving Mg unchanged. **Sabkha** dolomite (Fig. 2A) forms when storm surges force seawater landward over peritidal sediments in arid regions. Evaporation results in the precipitation of evaporite minerals. The increased density of the evaporation-concentrated brine causes it to flow downward and dolomitize underlying lime sediments. Thus, hydrodynamic head provides the hydrologic "pump." Sabkha dolomite is usually easy to identify, because it is always associated with supratidal sediments and contains many distinctive associated features, such as algal stromatolites, nodular anhydrites, and eolian interbeds. **Reflux** dolomite (Fig. 2B) forms beneath restricted lagoons and basins, where evaporation is intense and leads to brine concentration and evaporite precipitation. Downward and seaward density-driven flow of brine once again leads to dolomitization of underlying sediment. However, in this case, the underlying sediment need not be supratidal, but can represent any depositional facies.

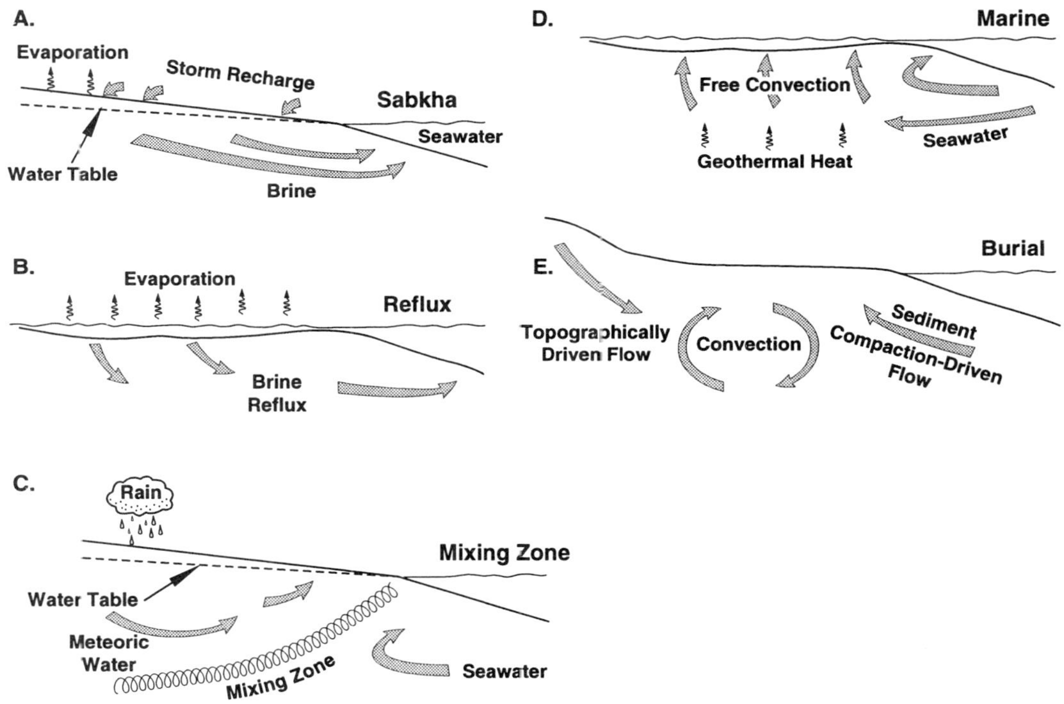

Modified after Land (1986) and Morse and Mackenzie (1990)

Figure 2. Illustration of the hydrologic environments which produce (A) sabkha, (B) reflux, (C) marine-meteoric mixing zone, (D) marine, and (E) burial dolomite. The hydrology of the dolomitizing environment controls the shape, size, and location of the dolomite body produced. See text for discussion.

The marine-meteoric **mixing zone** model (Fig. 2C) is based on the idea that mixing meteoric water with seawater can produce a fluid that is undersaturated with respect to calcite and aragonite and supersaturated with respect to dolomite. The Mg is supplied by seawater and hydrodynamic head provides the hydrologic pump. Although the model has been widely invoked to explain ancient dolomitized sequences which lack evaporites, it is flawed in several ways. The absence of dolomite in all but a few modern mixing zones and the fact that seawater alone is more supersaturated with respect to dolomite than any meteoric water - seawater mixture are the two most commonly cited problems. As a result, increasing attention has been given to dolomitization by normal seawater.

Strong evidence for dolomitization by seawater has been presented in many recent papers. **Marine** dolomite (Fig. 2D) derives its Mg directly from seawater. The Mg is delivered either by tidal pumping or by inflow and buoyant rise of geothermally heated seawater at continental margins. **Burial** dolomite (Fig. 2E) forms in the subsurface after lithification of lime sediments. As temperature increases, the Mg/Ca ratio required to produce dolomite decreases and many subsurface waters can become dolomitizing solutions. The most-cited fluid sources for burial dolomitization are: (1) Mg-rich residual evaporite brines, (2) modified seawater, and (3) shale compaction waters. The water can be delivered to the site of dolomitization through porous aquifers or along faults by a number of transport mechanisms, including sediment compaction, thermal convection, and topographically or tectonically driven flow.

DOLOMITE PETROGRAPHY

Although often critical to reservoir development and porosity distribution, dolomite crystal size and texture are rarely definitive of dolomitizing environment. Nonetheless, crystal size and texture can be helpful in identifying a few specific types of occurrence. Therefore, before moving on to the geochemistry section of this manual, we will briefly examine dolomite petrography.

Microcrystalline texture in dolomite (particularly crystal size $<10\mu m$) is believed to be caused by a high degree of supersaturation with respect to dolomite, which overcomes kinetic barriers to nucleation and results in rapid precipitation of dolomite at many nucleation sites (Sibley, 1991). Microcrystalline textures are most commonly associated with sabkha dolomites (Moore, 1989) and rarely reported for other dolomitizing environments. Thus microcrystalline texture, particularly when associated with supratidal evaporites, is commonly indicative of early, nearsurface dolomitization.

Saddle dolomite (sometimes called baroque dolomite) is characterized by coarse crystallinity (crystal size from $\sim 150\ \mu m$ to several mm), curved crystal faces and cleavage, undulose extinction, and abundant fluid inclusions which give it a milky white to brown color. It occurs both as a void-filling cement in fractures and vugs and as a replacement of limestone (Radke and Mathis, 1980). Because saddle dolomite is commonly associated with hydrocarbons and hydrothermal minerals, has a "high temperature" geochemical signature, and, when replacive, transects bedding, stylolites, and formation boundaries, it is considered to be a late, high temperature phase.

Fluorescence microscopy can sometimes be used to visually recognize grain types and depositional textures in pervasively dolomitized carbonate rocks (Dravis and Yurewicz, 1985). Incident light fluorescence with a blue-light excitation filter apparently causes residual organic matter preserved in grains to fluoresce, allowing one to identify the depositional facies of the precursor sediment, even when dolomitization has nearly obliterated all textures. When successful, this technique provides information useful for interpreting dolomitization environment (e.g., open marine grainstones which are not likely to have been dolomitized by supratidal processes may be recognized). Unfortunately, fluorescence microscopy only works well in a minority of cases.

PART I:
INORGANIC GEOCHEMICAL TECHNIQUES

ANALYTICAL APPROACHES

Part I of this manual describes inorganic geochemical techniques that can be used to deduce the origin of a dolomite and establishes guidelines that geologists can follow to interpret dolomite geochemical data.

Our specific goal in studying dolomitization is to gain a better understanding of the factors that control a dolomite unit's extent and shape — to determine whether it is stratiform, structurally controlled, etc. The geochemical techniques that are most useful in attaining this goal are:

1. Produced formation water analyses
    a. Oxygen and hydrogen isotopic compositions
    b. Solute concentrations
2. Dolomite analyses
    a. Oxygen and carbon isotopic compositions
    b. Fluid inclusion homogenization temperatures and salinities
    c. Strontium isotopic compositions
    d. Trace element concentrations (particularly Fe and Mn)

The analytical techniques in (1.) shed light on whether dolomite has formed from present-day well or formation fluids. When analyzing a produced formation water, relevant questions one might ask are: "Did this water originate as expelled interstitial evaporite brine, invaded meteoric water, or seawater?"; and "Was the water ever involved in dolomitizing limestone?"

The analytical techniques in (2.) shed light on the physical and chemical environment in which a dolomite formed. When analyzing dolomite, relevant questions one might ask include: "At what temperature and depth did dolomitization occur?"; "What was the salinity of the fluid responsible for dolomitization?"; and "To what extent did organic matter maturation and degradation products and siliciclastic diagenesis contribute to the chemical composition of the dolomite?"

A list of the various techniques and their applications is presented in Table 1.

In this part of the manual, we introduce and discuss the geochemical techniques that are most commonly used in dolomite studies and provide the explorationist with advice on how best to apply the techniques, interpret the data, and recognize and avoid some common pitfalls and misconceptions that one commonly encounters.

When beginning a study, the techniques presented in Part I should be reviewed before samples are selected for analysis. The explorationist should decide which analyses he or she wishes to perform and should then read the appropriate section of the text and any appropriate articles cited there. When the geochemical analyses are complete, geochemical data can be integrated with the results of stratigraphic, structural, petrographic, and sedimentary facies analysis and compared to the case studies presented in Part II of the manual.

Table 1. Analytical Techniques and the Questions Concerning Dolomite Models Which They Address

ANALYTICAL TECHNIQUE			Relevant Questions Addressed	Pages
Formation Fluid	Isotopes	$\delta^{18}O$	1. Meteoric, marine or evaporative brine? 2. History of interaction with carbonates?	9-12
		δD	1. Meteoric or marine fluid?	9-12
	Solutes	Br and C. F. vs. Chlorine plots	1. Meteoric, marine or evaporative brine? 2. Is the fluid the product of fluid mixing? 3. Has the fluid dissolved evaporites? 4. Is the fluid chemically capable of dolomitization?	12-18
		Ca/Mg vs. T	1. Has the formation fluid formed dolomite (Lost Mg)? 2. Is the fluid still capable of dolomitization?	16-18
Dolomite	Isotopes	$\delta^{18}O$	1. Low or high temperature of dolomitization? 2. Meteoric water overprint? 3. Recrystallization?	19-32
		$\delta^{13}C$	1. Dolomitization after methane oxidation, permentation, biodegradation, thermal decarboxylation?	33-38
	Fluid Inclusions	Fluorescence	1. Dolomitization in an oil migration pathway? 2. Maturity (API Gravity) of hydrocarbon?	40-41
		Th	1. Temperature of dolomitization? 2. Age of dolomitization relative to basin's thermal history?	41-43
		Tm	1. Salinity of dolomitizing fluid (meteoric, marine, evaporative brine, etc.)?	43-45
		Th vs. Tm plots	1. Identification of altered fluid inclusions. 2. Basin history (subsidence, uplift, thermal event) subsequent to dolomitization?	45-48
	Strontium Isotopes	$^{87}Sr/^{86}Sr$	1. Marine continental or deep burial environments of dolomitization? 2. Age of marine fluid that formed dolomite?	49-52
	Trace Elements	Fe, Mn	1. Marine or burial dolomitization environment? 2. Oxidation-reduction potential and H_2S content of dolomite-forming fluid?	53-55
		CL-zonation	1. Fluctuating pore water chemistry?	55-57

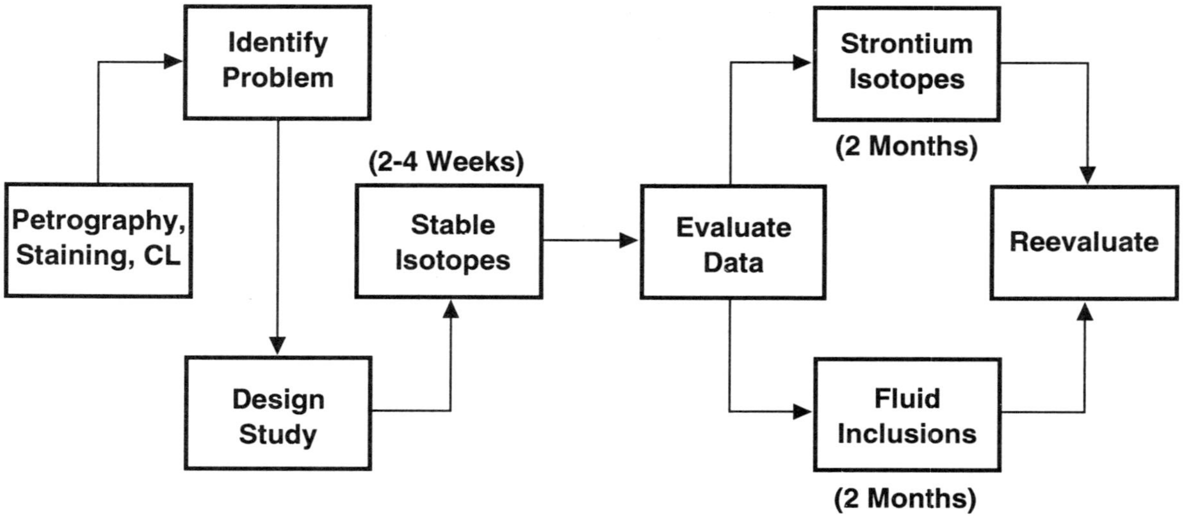

Figure 3. Flow chart for proposed geochemical study illustrating typical amounts of time required to obtain different types of geochemical data from commercial laboratories.

FORMATION WATER ANALYSES

Many types of fluid within sedimentary basins are capable of forming dolomite on a regional scale. The major fluid types that must be considered are evaporite-related brines[1], invaded meteoric waters, and seawater.

The deeper parts of many basins are now pervaded by brines that originated as interstitial fluids in evaporite-bearing sections. The brines can move both upward and downward from their source beds. An example of upward movement of brine is the Louann Salt (Jurassic-aged) interstitial fluids in the Gulf Coast that moved upward into Cretaceous reservoirs. An example of downward movement occurs in the Michigan Basin where Silurian-aged Salina Group brines migrated downward into Ordovician reservoirs. Brines such as these, which were expelled from their evaporite source beds, are referred to as "residual evaporite brines" (Carpenter, 1978; Hanor, 1987) because they are the fluids which remain after precipitation of soluble salts. Subsurface brines may also be generated by the dissolution of halite, however such fluids are generally incapable of forming massive quantities of dolomite due to insufficient magnesium content (see Land and Prezbindowski, 1981 and 1985).

The next two sections illustrate a strategy for investigating formation fluids, especially brines, in order to determine their origin and capacity to form dolomite.

Stable isotopes are treated first in our discussion of formation fluids. The reader should refer to Appendix I for a description of the notation and standards used in reporting isotope data before continuing in this chapter. After treatment of stable isotopes, we explain the importance of evaluating the solutes dissolved in formation water.

STABLE ISOTOPES

Several stable isotopes in water are important: ^{18}O, ^{16}O, ^{2}H (or D for deuterium), and ^{1}H. The isotopic composition of water is expressed as $\delta^{18}O$ and δD (see Appendix II for the definition of this notation). We refer to water as "light" when it is depleted in the heavier isotopes ^{18}O and D and "heavy" when it is enriched in these isotopes. The stable isotopes of oxygen and hydrogen can be used to test whether a produced water sample is derived from meteoric water, seawater, or evaporites, and whether this water has reacted extensively with carbonates.

Three general types of water are evident on a standard $\delta^{18}O$-δD crossplot: (1) meteoric water, (2) seawater or evaporated seawater and (3) water whose isotopic composition has been altered by rock-water reactions. Each of these cases is discussed below.

(1) There is a strong covariance between $\delta^{18}O$ and δD of rain and snow from around the world (Fig. 4) because of the following mechanisms: During evaporation of seawater, lighter molecules of H_2O (containing a greater proportion of ^{16}O and ^{1}H than seawater) are selectively evaporated from the ocean so that the ratio of heavy to light molecules is lower in rain than in the oceans. Meteoric water becomes isotopically more depleted in ^{18}O and D

[1] The term "saline formation water" is reserved for fluids with salinities of between 35,000 mg/l (close to the salinity of seawater) and 100,000 mg/l, and the term "brine" signifies a salinity of greater than 100,000 mg/l.

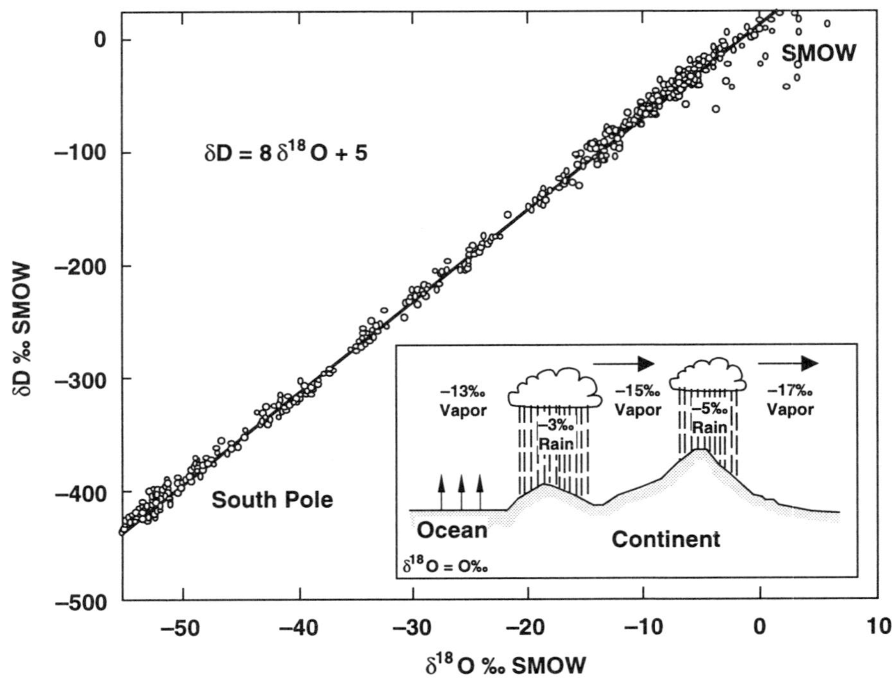

Hoefs (1980)

Figure 4. Waters of meteoric origin are easily identified using stable isotopes because of the high degree of correlation between oxygen and hydrogen isotopes in meteoric waters worldwide. During evaporation of seawater, molecules of water containing a smaller proportion of ^{18}O and D than seawater are selectively evaporated from the ocean. Water vapor rising over the oceans cools until the dew point is reached. During the outraining of clouds, the rain is enriched in ^{18}O and D and the remaining moisture in the cloud is further depleted in ^{18}O and D (see inset). As an air mass moves poleward, inland or higher in elevation, and cools, the rain produced becomes increasingly depleted in both of these isotopes, leading to the spread in values observed along the Meteoric Water Line.

with increasing latitude, altitude and distance inland over the continental landmasses (Fig. 5) because of the general pattern of poleward and onshore circulation and cooling. The "lightening" of oxygen and hydrogen isotope ratios (i.e., the increase in the proportion of the lighter isotopes, ^{16}O and ^{1}H) with increasing latitude, altitude and distance inland is caused by multiple evaporation-reprecipitation events and increased liquid-vapor fractionation events with decreasing temperatures as the tropical air masses move away from the equator and inland. High-latitude, continental groundwaters are depleted in ^{18}O and D and plot down and to the left on the Meteoric Water Line (MWL), far from tropical, coastal rainwater, which plots near to seawater (Standard Mean Ocean Water or SMOW on Fig. 4). Tropical rainwater most often influences carbonate diagenesis during the early history of a carbonate unit ("eogenesis" of Choquette and Pray, 1970), whereas high-latitude continental rainwater most often influences carbonate diagenesis long after lithification and tectonic uplift into the zone of meteoric circulation ("telogenesis" of Choquette and Pray, 1970).

(2) Seawater, the second important fluid in dolomitization processes, plots at the origin (SMOW in Fig. 4) of the $\delta^{18}O$-δD diagram. When it is evaporated, for example in a sabkha setting, its isotopic composition moves on the path shown in Figure 6 as lines H.H. (high humidity)

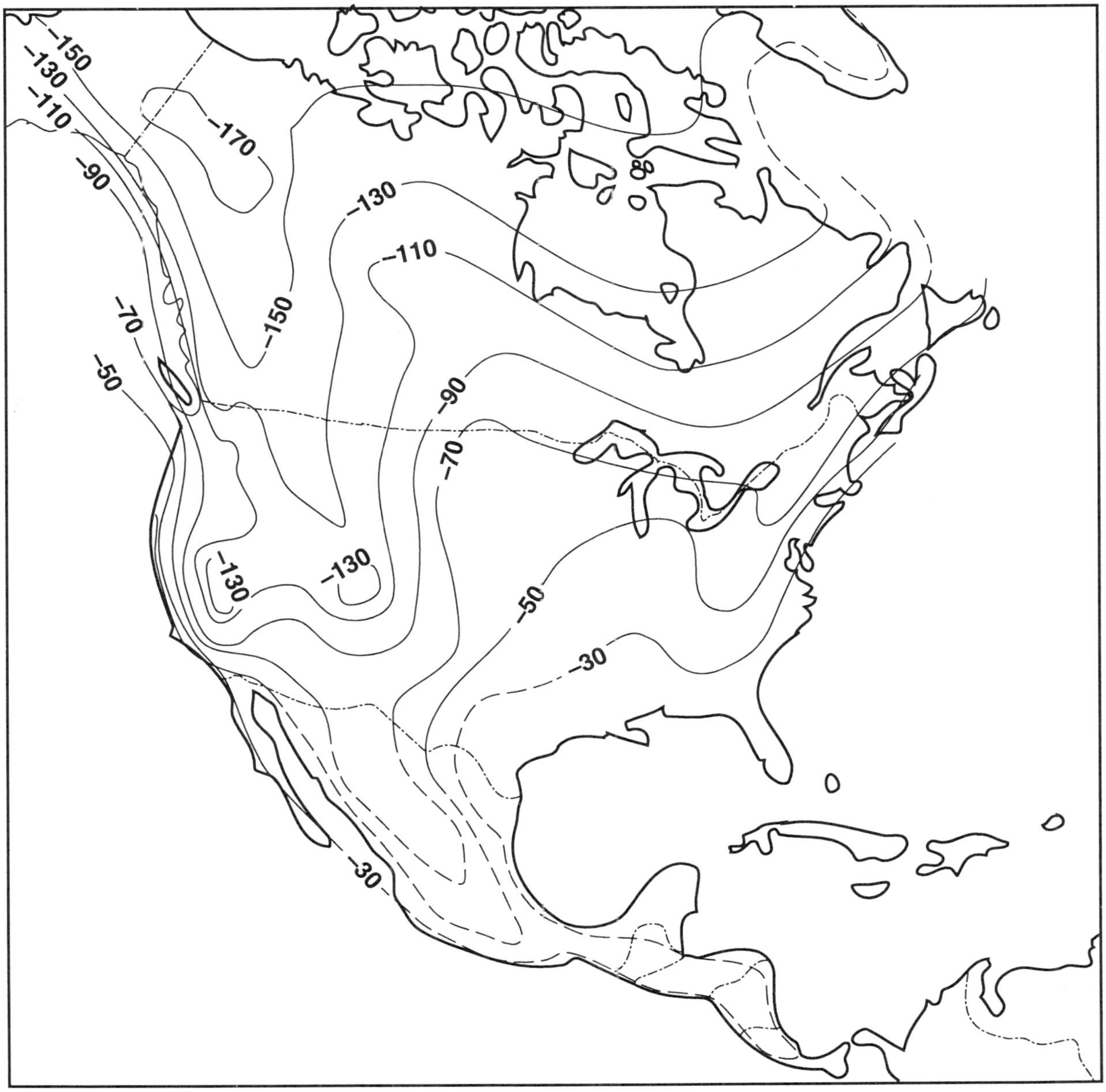

Contours = $\delta D_{Rain/Snow}$ ‰ SMOW

Taylor (1974)

Figure 5. The oxygen and hydrogen isotopic compositions of rain and snow vary systematically with respect to latitude, elevation and distance inland. This contour map of the δD of precipitation in North America shows a concentration of isotopically light precipitation at higher latitudes along the Rocky Mountains. A similar map can be constructed from $\delta^{18}O$ data.

and L.H. (low humidity). The evaporation path can plot as a "hooked" trajectory (Line E in Fig. 7). Trajectory E is caused by the sequestering of ^{18}O by hydrous salts at extreme ionic strengths, which causes the isotopic composition of free water to become lighter with increasing salinity. Most residual halite brines fall on or near this trajectory (Knauth and Beeunas, 1986).

(3) Water of any kind can be heated during burial in a subsiding basin and will dissolve and reprecipitate carbonate minerals at elevated temperatures ("mesogenesis" of Choquette and Pray, 1970). When this occurs, some of the ^{18}O in the carbonate minerals exchanges with some of the ^{16}O in the formation water. This leads to a depletion of ^{18}O in the mineral, and its enrichment in the water (Land, 1980). Formation waters that plot to the right of the Meteoric Water Line along a positive $\delta^{18}O$ trajectory at a constant δD have most likely undergone extensive reaction at elevated temperatures with carbonate minerals (Line W.R.I. in Fig. 6). This means that the end products of evaporation and rock-water interaction can have similar isotopic compositions, i.e., both processes lead to ^{18}O-enriched dolomite.

Any or all of these three basic water types can commingle in the subsurface, the isotopic composition of commingled waters being the weighted average of the isotopic compositions of the components. For example, brines expelled from evaporites that emerge from the deep subsurface and mix with waters of meteoric derivation plot on mixing lines such as those shown in Fig. 7.

By way of illustration, consider an isotope data set from the Alberta Basin, western Canada (Fig. 8). Plotted in Fig. 8 are the oxygen and hydrogen isotopic compositions of surface samples of meteoric water collected from lakes, rivers and shallow wells in Alberta and northeastern British Columbia (solid circles), present-day seawater (SMOW), and formation waters from the deep Alberta Basin (open squares). Using this plot, one can determine that the formation waters are a mixture of meteoric water and brines which formed by the evaporation of seawater and were perhaps further enriched in ^{18}O through extensive rock-water interaction (primarily subsurface dolomitization) at elevated temperatures.

SOLUTES

Saline formation waters are often analyzed by oil companies for their major, minor and trace cation and anion concentrations to help deduce their origin and evolution. We have two general purposes in looking at dissolved constituents: (1) to identify the origin of the fluid, and (2) to determine whether it had a history of dolomitization.

The concentration of solutes in seawater changes predictably during evaporation (Fig. 9). Because most formation water originates as seawater, we can use the concentrations of particular solutes in primary or unaltered seawater as baselines to track formation water history. By comparing the change in concentration of elements which are precipitated out in salts during the course of seawater evaporation with the change in concentration of bromine, which is not precipitated out, we can decipher the evaporative history of a seawater-derived brine, and recognize brines which were altered by dilution, the addition of soluble salts, or by simple evaporation-precipitation.

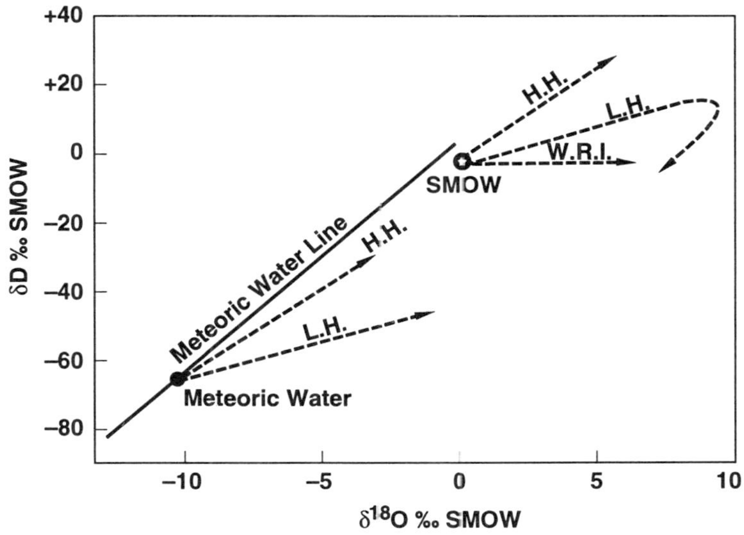

(From various authors)

Figure 6. Shown are the isotopic compositional trajectories for seawater and high-latitude meteoric water which have been evaporated at low humidity ("L.H.") and at high humidity ("H.H."). Note that any water, including meteoric water, will follow a predictable trajectory when subjected to evaporation. The slope of the evaporation trajectory varies with humidity, being steeper for high humidity conditions (H.H.). The change in the composition of seawater which has been influenced by water-rock interaction is shown as "W.R.I.".

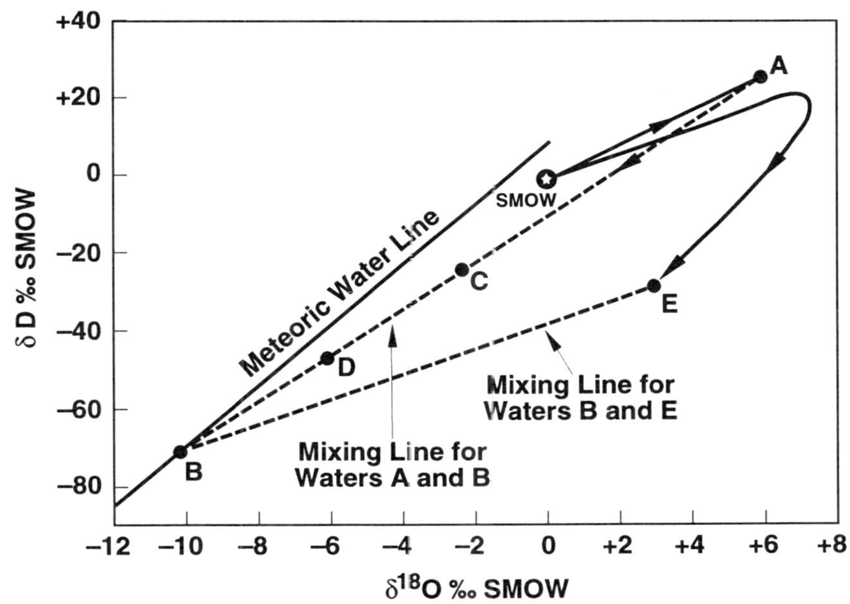

Knauth and Beeunas (1986)

Figure 7. Both marine evaporite brines and fluids which are products of water-rock interaction, when mixed with meteoric water or seawater, will lie to the right of the Meteoric Water Line. Typically, an evaporated brine is an end point on a mixing trend. In this example, a marine fluid starts with a composition of SMOW and then evaporates to Point A. At a much later time in its geologic history, as this brine mixes with recharging meteoric water (Point B), the composition of the brine will be displaced from Point A toward Point B, the amount of displacement depending on the proportions (50:50 meteoric water:brine mixture plots at Point C, and a 75:25 mixture at Point D). If this brine were an interstitial fluid from halite-bearing evaporites, its starting composition would have been closer to Point E. Further evaporation of any of these fluid mixtures could drive their isotopic compositions along a line trending parallel to the line extending between SMOW and Point A.

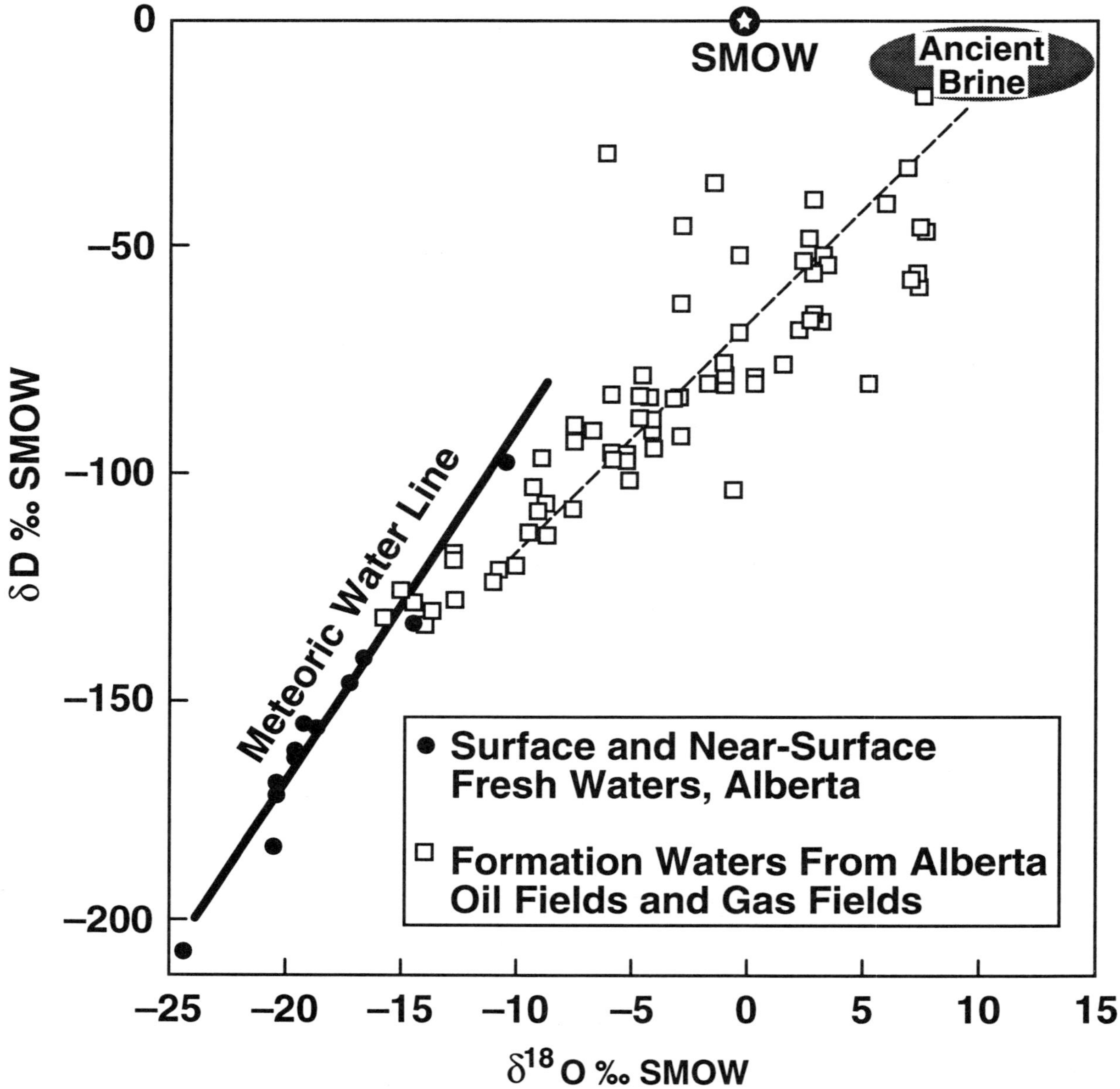

Hitchon and Friedman (1969)

Figure 8. Most Alberta Basin formation waters plot on a mixing line between ancient (Devonian) brines from trajectories L.H. and W.R.I. in Fig. 6 (upper right part of this figure) and present-day meteoric water (lower left part of this figure). The formation waters which plot closer to the Meteoric Water Line are from Cretaceous and younger sediments.

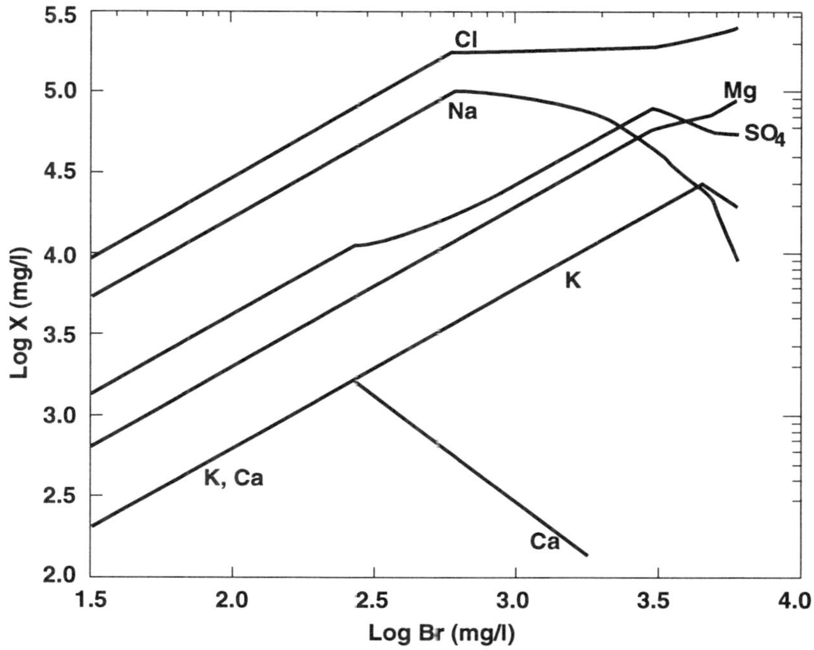

Carpenter (1978)

Figure 9. The concentrations of solutes in seawater change predictably during evaporation. Ca removal commences at the onset of gypsum precipitation, followed by Na and Cl removal by halite precipitation. Note that Cl continues to increase in concentration in seawater even in the halite stability field, but at a slower rate than before. Also, it is not until the stage where bitterns form that we see removal of Mg (as Mg-sulfates) and K (as carnallite). Br is conserved in the brine throughout the sequence. The Cl trend line is of special significance (see text and Figures 10 and 11).

Crossplots of conservative ions versus non-conservative ions are very useful for tracing the origin and evolution of formation waters. The most commonly crossplotted ions are Br, which is conserved during the evaporation of seawater and subsequent rock/water interaction, and Cl, which is not conserved. Because oil field brines are not routinely analyzed for Br, however, a different conservative parameter must often be used (if Br data are available, then the Br-Cl crossplot is still the recommended evaluation technique).

Carpenter (1978) shows that the quotient [Ca+Mg+Sr-SO_4-HCO_3 meq/l]/[Br mg/l] equals approximately 1.0 for seawater. During evaporation of seawater, this ratio remains constant up to the point where carnallite begins to precipitate (well beyond initial halite precipitation). This relationship persists because most exchange reactions that affect ions in the numerator are exchange reactions that change the amount of individual ions (i.e., Ca, Mg, Sr, SO_4 and HCO_3) but not their sum. Because the ratio is nearly 1.0, Br (mg/l) and Ca+Mg+Sr-SO_4-HCO_3 (meq/l) can be used interchangeably in crossplots and calculations.

Both Cl and the ($Ca+Mg+Sr-SO_4-HCO_3$) parameter[2] are conserved in solution and increase in a 1:1 ratio during the evaporative concentration of seawater until halite saturation is reached. Once halite precipitation begins, Cl is incorporated in the mineral phase and accumulates more slowly in solution. Figure 10 is a crossplot of log $Ca+Mg+Sr-SO_4-HCO_3$ (meq/l), the conservative parameter, versus log Cl (mg/l), the non-conservative parameter. Note that at the point of halite precipitation where Cl begins to be removed, the slope of the seawater evaporation line changes abruptly. Water that plots on the seawater evaporation line to the left of the break in slope is simply evaporation-concentrated seawater. Water that plots on the line to the right of the break in slope is original, evaporation-concentrated interstitial water from halite-bearing evaporites ("interstitial halite brine"). Water that plots below the line probably formed by mixing of an interstitial halite brine with a water of low salinity, for example, seawater or fresh water (Fig. 11A). Brine that plots above the seawater evaporation line in Fig. 11B has had a Na-Cl brine component introduced by dissolution of halite.

Conservative-ion versus Cl plots may be used to determine whether a particular formation water originated as seawater, evaporated seawater, interstitial halite brine, halite dissolution brine, or to determine whether it is of mixed parentage. An example of this application to the subsurface Texas Gulf Coast is shown in Figure 12.

This is not the end of the analysis. Brine evaporated past the point of halite saturation is strongly enriched in magnesium, and is thus a potent dolomitizing agent. If an interstitial halite brine is identified from the conservative/non-conservative ion crossplot, one should suspect that it may have produced dolomite somewhere in the basin during burial diagenesis. To further test whether a present-day formation fluid of evaporite-brine parentage has participated in dolomitization reactions, we can determine its Ca/Mg activity ratio at reservoir temperature and compare the ratio to those at empirically-determined calcite and dolomite stability field boundaries (Carpenter, 1980), and to its initial Ca/Mg activity ratio (i.e., that of seawater evaporated to halite saturation). If the Ca/Mg activity ratio has increased from that of interstitial halite brine, it is likely that the formation fluid has been involved in dolomitizing limestone (Fig. 13).

For example, in Figure 13, an unaltered interstitial halite brine (identified using Fig. 9 or 10) should plot to the right of the halite saturation line, far within the dolomite stability field. If, however, it plots at or near equilibrium with calcite and dolomite, then it has lost much of its magnesium to dolomitization. The simplest explanation for such a fluid is that a Mg-enriched, interstitial halite brine generated in a highly evaporitic section migrated into limestone during burial, mixed with interstitial fluids, and dolomitized the limestone. Once the Ca/Mg ratio of such a brine is raised to a value in equilibrium with calcite plus dolomite, dolomitization will cease and, although the brine may migrate farther into additional limestone, no more dolomite will be produced.

In summary, isotope and chemical data sets can be used to unravel the origin and migration of subsurface fluids. A complete formation water data set, which includes both stable isotopes and solute concentrations, would allow evaluation of whether: (1) the fluid was derived from seawater, meteoric water, evaporites, or any mixture of the three (through the use of stable isotope and conservative ions versus chloride plots), (2) whether the fluid interacted with carbonates (from stable isotope plots), and, finally, (3) whether the fluid participated in the dolomitization of limestone (by means of the Ca/Mg activity ratio versus temperature plot).

[2] [$Ca+Mg+Sr-SO_4-HCO_3$ meq/l] has recently been called the "Carpenter Function", or "C.F."

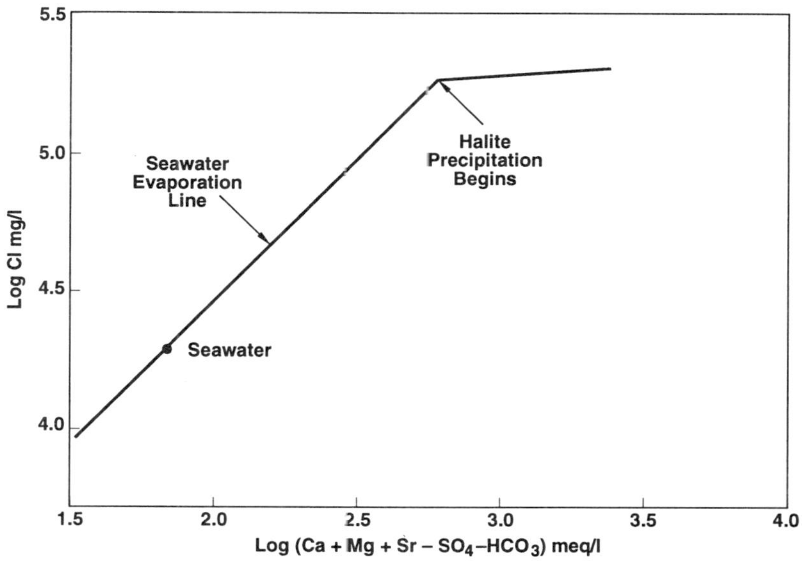

Carpenter (1978)

Figure 10. A convenient method for tracking the history of a marine fluid is through the crossplot of a non-conservative ion such as Cl, against a conservative parameter, in this case (Ca+Mg+Sr-SO_4-HCO_3). Both are conserved in the fluid up to the point of halite precipitation, where Cl begins to be removed. Waters that plot to the right of the break in slope where halite begins to precipitate are original, evaporation-concentrated marine brines from halite-bearing evaporites. Note the shape of this trend line is identical in shape to the Cl vs Br trend presented in Fig. 9. The Br vs Cl plot in Fig. 9 is the more conventional method for tracking a fluid's composition, however the plot presented here is acceptable when Br data are not available.

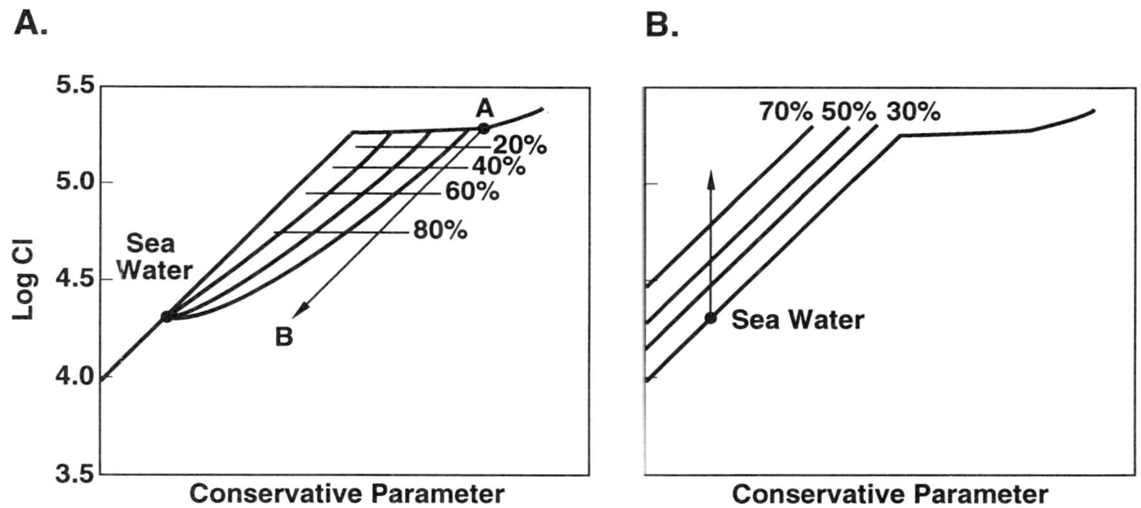

Carpenter (1978)

Carpenter (1978)

Figure 11. (**A**) Waters may plot off the evaporation trend line introduced in Fig. 10. In Fig. 11A, they plot below the line and are products of mixing an evaporation-concentrated, marine brine (Point A) with a lower-salinity water (Point B). The addition of normal seawater to a brine of composition A causes the brine composition to move along curved trends which converge at seawater. The percentage of seawater in the mixture is indicated on the horizontal isopleths (lines of constant composition). (**B**) Waters that plot above the evaporation trend line are products of halite dissolution. The isopleths indicate the percentage of chloride in seawater-derived brine which has been contributed by dissolving halite. Waters of this type have a reduced potential for dolomitization (Land and Prezbindowski, 1981 and 1985).

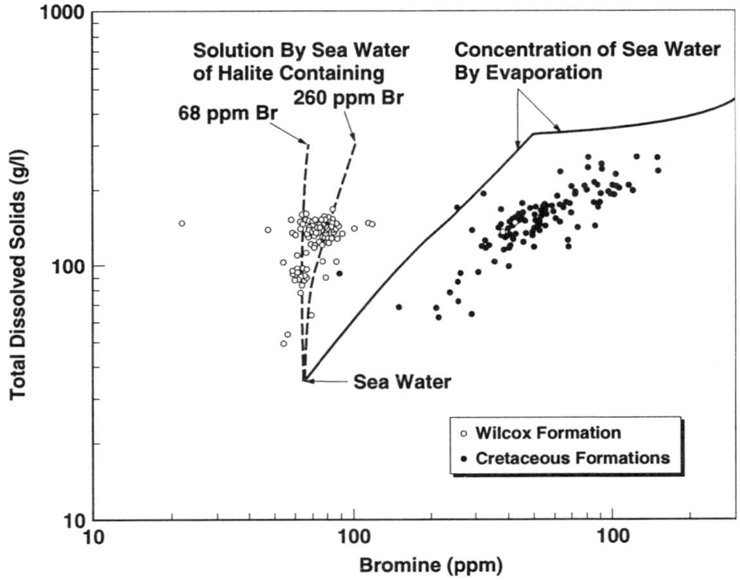

Rittenhouse (1967)

Figure 12. Br and Total Dissolved Solids (TDS) data from the Alabama-Mississippi Gulf Coast Region are shown for the Eocene Wilcox Fm. and Cretaceous formations. The plot illustrates that Wilcox samples are enriched with the products of halite dissolution because of their proximity to salt domes which penetrate the Wilcox Fm., whereas brines in the Cretaceous formations are free of dissolved halite.

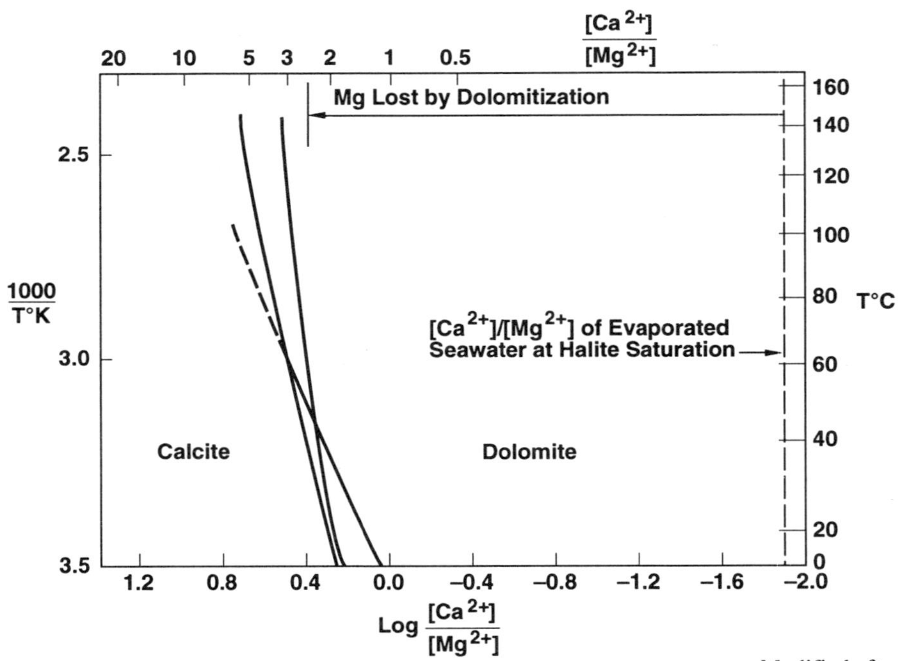

Modified after Carpenter (1980)

Figure 13. To test whether a residual evaporite brine has acted as a dolomitizing fluid, formation water Ca/Mg ratios and fluid temperatures can be compared to empirically-determined calcite/dolomite stability field boundaries (Carpenter, 1980). A brine that has a lowered Ca/Mg ratio due to precipitation of gypsum, anhydrite or other evaporite minerals, but has not been involved in dolomitization, will plot well within the dolomite stability field, indicating this fluid still has a high potential for dolomitization. On the other hand, a fluid that has already formed large amounts of dolomite will be depleted in magnesium and will plot on or near the transition between calcite stability and dolomite stability.

OXYGEN ISOTOPES IN DOLOMITE

Both elevated temperature and the "meteoric effect" (i.e., decreasing $\delta^{18}O$ of rainfall and fresh groundwater with increasing latitude, altitude, and distance from the ocean) drive the oxygen isotopic composition of diagenetic carbonate negative (Fig. 14; Taylor, 1974). However, because dolomite rarely forms from meteoric water, significant ^{13}O depletion in dolomite is usually taken to indicate precipitation at elevated temperature. This temperature-$\delta^{18}O$ relationship is explained next.

Carbonate minerals always concentrate ^{18}O relative to the fluid from which they were deposited. This means that the ratio of ^{18}O to ^{16}O in the mineral is greater than that in the parent fluid. Segregation of ^{18}O and ^{16}O between the parent fluid and mineral (known as fractionation) is less and less efficient at increasing temperatures, meaning that the $^{18}O/^{16}O$ ratios of the mineral and the parent fluid become closer together at higher crystallization temperatures. To illustrate, consider how differences between calcite and standard mean ocean water (SMOW) diminish with higher temperatures, and how this affects the δ value of calcite (cc):

	$[^{18}O/^{16}O_{cc}]/[^{18}O/^{16}O_{water}]$	$\delta^{18}O_{water}$ SMOW	$\delta^{18}O_{cc}$ SMOW	$\delta^{18}O_{cc}$ PDB
@25°C	1.0289	0	+28.5	-2.3
@50°C	1.0243	0	+24.0	-6.7
@100°C	1.0173	0	+17.2	-13.3
@500°C	1.0018	0	+1.8	-28.2

It is by virtue of this temperature dependence that oxygen isotope ratios in carbonate minerals serve as useful geothermometers. The following equation from Friedman and O'Neil (1977) is often used for modelling the temperature of dolomitization:

$$\underset{\text{(CALCULATED)}}{[3.2 * 10^{6}\mathbf{T}(^{\circ}\mathbf{K})^{-2}]} - 1.5 = \underset{\text{(MEASURED)}}{\delta^{18}O_{dol}} - \underset{\text{(ESTIMATED)}}{\delta^{18}\mathbf{O}_{\mathbf{water}}} \quad (1)$$

This equation has two unknown values, the oxygen isotopic composition of water ($\delta^{18}O_{water}$) and temperature (T), and one known value, the oxygen isotopic composition of dolomite ($\delta^{18}O_{dol}$). Because it is T that we are normally trying to calculate, and only $\delta^{18}O_{dol}$ is measured, it becomes necessary to constrain the value of $\delta^{18}O_{water}$. One approach for estimating $\delta^{18}O_{water}$ is to use the fluid inclusion homogenization temperature (Th, the temperature at which a two-phase inclusion — liquid and its vapor — becomes one phase, provides an estimate of the minimum temperature of crystallization for the host crystal) from a sample on which both fluid inclusion and isotope data have been collected. Th can then be substituted for T in the isotope temperature equation (1) to calculate the isotopic composition of the dolomitizing fluid. Then, with this calculated composition of $\delta^{18}O_{water}$, dolomitization temperatures can be calculated using $\delta^{18}O_{dol}$ for other samples that lack fluid inclusion data. A common alternative approach is to estimate $\delta^{18}O_{water}$ based on geologic reasoning

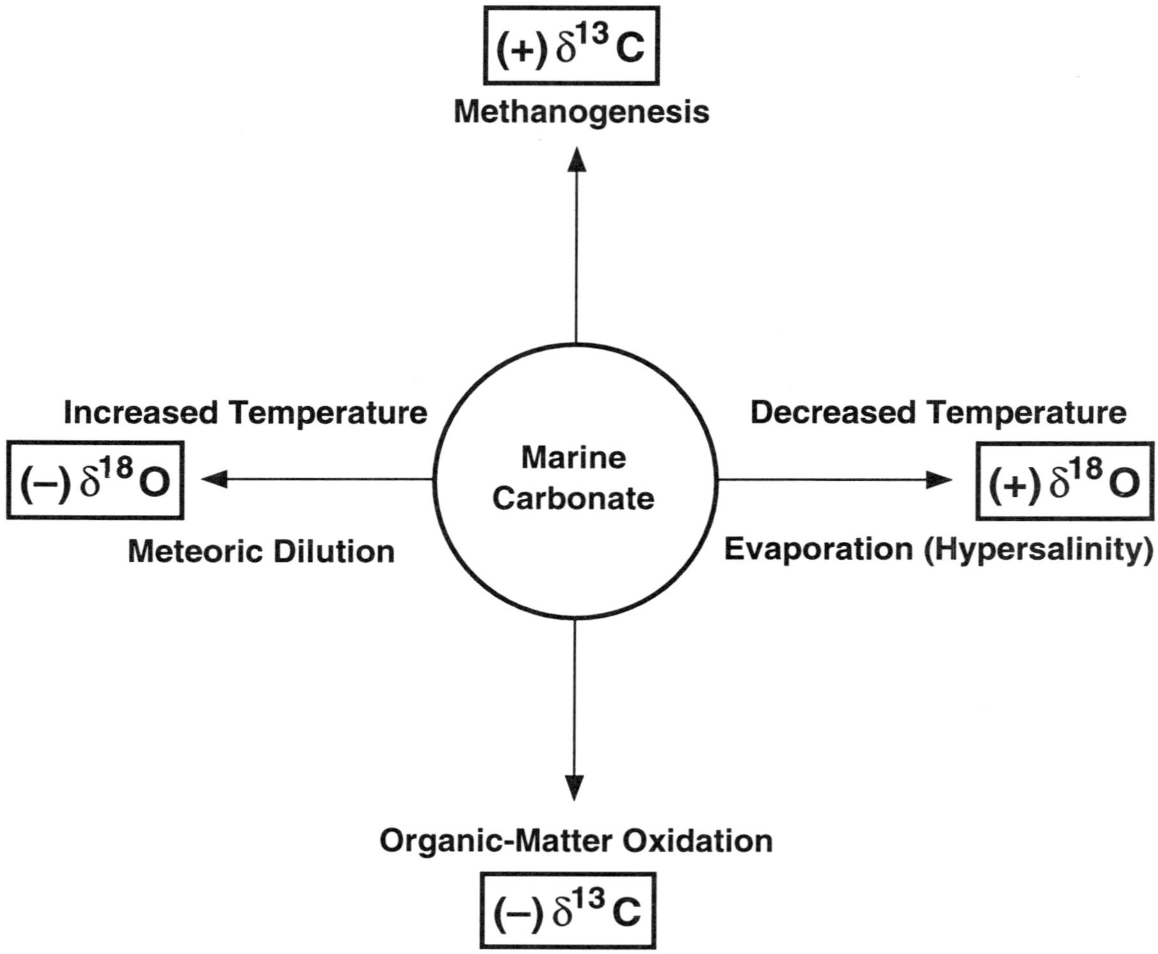

Modified after Lohmann (1988a)

Figure 14. Summary of the effects of temperature, evaporation, and dilution of the dolomitizing fluid, plus input of CO_2 derived from methanogenesis and organic matter oxidation on the isotopic composition of dolomite. By comparing dolomite isotope ratios to that of a marine carbonate baseline (central part of diagram), we can gain a better understanding of the environment in which dolomite formed.

and to substitute the estimated value into Equation (1) to calculate T. However, the error in calculated T can be very large if the estimated value of $\delta^{18}O_{water}$ is uncertain.

A more conservative approach is to avoid assumptions about $\delta^{18}O_{water}$. Because of the uncertainty in $\delta^{18}O_{water}$, the best we can accomplish in most cases is general distinction between high-temperature (burial) dolomite and low-temperature (near-surface) dolomite. This can be done empirically by using two approaches. In the first approach, dolomites of unknown origin are compared to marine carbonates precipitated in equilibrium with seawater at surface temperatures. In the second, dolomites of unknown origin are compared to other dolomites whose origin has been established by independent means, such as fluid inclusions, or through an association with ore deposits.

APPROACH #1. COMPARISON TO MARINE CARBONATES

To distinguish high and low temperature dolomites, one can compare a dolomite oxygen isotopic composition to the oxygen isotopic composition of unaltered primary marine calcium carbonate. Marine carbonate sediments, shell material, and cements originally form as low-magnesium calcite, high-magnesium calcite, or aragonite at or near isotopic equilibrium with seawater at earth surface temperatures. Although marine carbonate ($CaCO_3$) and early dolomite ($CaMg(CO_3)_2$) have different mineralogies, their isotopic compositions are related to one another in a predictable fashion. Therefore, the isotopic composition of marine carbonate can provide a useful baseline to which one can compare the isotopic composition of dolomite.

This marine baseline shifts with geologic age. Studies have shown that there have been secular changes in the initial isotopic composition of marine carbonate material. Whether this is due to changes in ocean temperature or to changes in the isotopic composition of seawater is still the subject of debate. However, the isotopic compositions of the shells and skeletons of marine invertebrates (brachiopods, belemnites, molluscs, and foraminifera) and marine carbonate cements have been measured and from this data set the isotopic composition of marine carbonate material can be estimated for each geological period.

Diagenesis, both in meteoric water and at higher temperature during burial, resets $\delta^{18}O$ to more negative values. This is generally, but not always, accompanied by a smaller change toward lighter $\delta^{13}C$ values (James and Choquette, 1983). As a result, the most-enriched $\delta^{18}O$ and $\delta^{13}C$ values in a marine carbonate data set are commonly considered to be the best estimates of original marine carbonate isotopic composition.

Two approaches have been used to estimate original marine isotope values. Lohmann and coworkers have restricted their studies primarily to marine carbonate cements. Using diagenetic trends derived from analysis of partially altered samples, they have extrapolated "least-altered" marine carbonate values (Fig. 15). Their method is described in detail in Appendix II. The results of many such studies are summarized in Fig. 16 (from Lohmann and Walker, 1989).

Researchers investigating the isotopic compositions of marine invertebrate shells and skeletons have taken a different approach. Workers in this field have developed a series of petrographic, mineralogic, and trace element criteria for identifying "least-altered" samples. These criteria, which are independent of isotopic evidence, are used to eliminate samples showing signs of diagenetic overprint *before* measuring isotope ratios. Popp, et al. (1986a) present $d^{18}O$ and $d^{13}C$ data for "least-altered" portions of Paleozoic brachiopod shells that were screened for diagenetic effects using cathodoluminescence (CL) and trace element analysis. Their brachiopod data are compared to Lohmann and Walker's marine cement data in Fig. 17. The two data sets agree in general, but there are several intervals in which cement and fossil data diverge, leading one to conclude that using a single point to represent the value for pristine marine carbonate for each period, as shown in Fig. 16, may be overly simplistic.

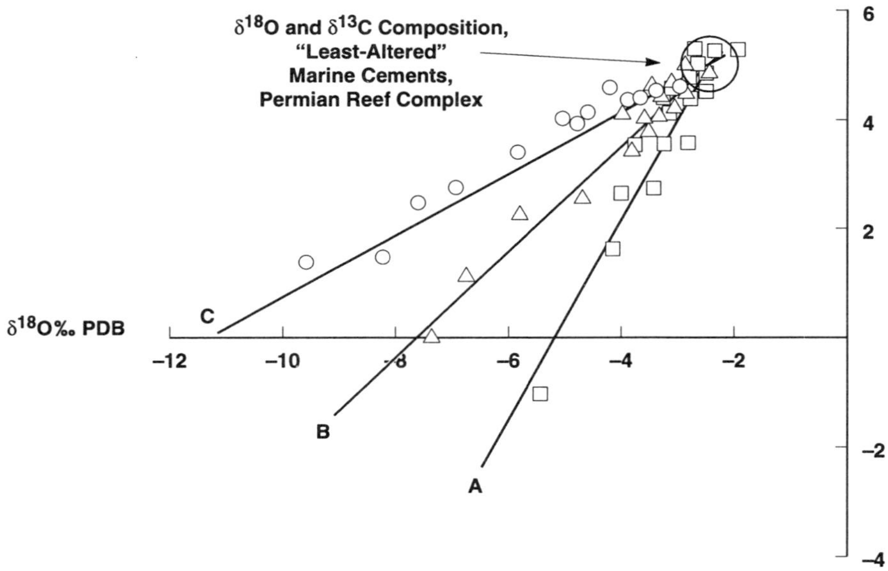

Modified after Given and Lohmann (1985)

Figure 15. Multiple isotopic analyses of former aragonite marine cement from Permian Reef Complex, West Texas and New Mexico, define linear trends for each of 3 sampled sites (lines A, B, and C). Trends are covariant in $\delta^{18}O$ and $\delta^{13}C$ and represent mixtures with different proportions of nonluminescent (marine cement) and luminescent (later diagenetic) calcite. $\delta^{18}O$ and $\delta^{13}C$ become more positive as the percentage of luminescent calcite decreases. The three trends converge on a single value, -2.5‰ $\delta^{18}O$ and +5.3‰ $\delta^{13}C$, which is interpreted to be the composition of the nonluminescent calcite (i.e., the marine cement). The large circle represents the error inherent in the interpretation.

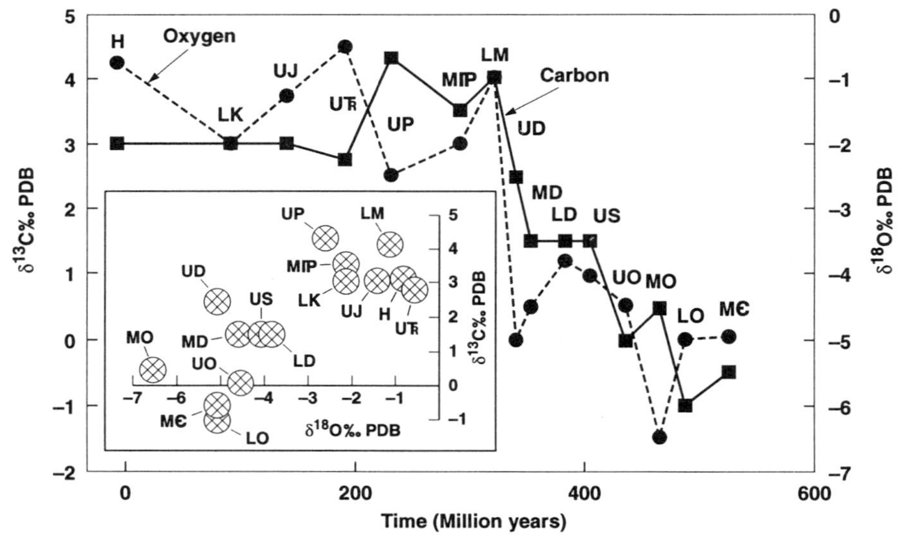

Modified after Lohmann (1988) and Lohmann and Walker (1989)

Figure 16. Secular changes in $\delta^{18}O$ and $\delta^{13}C$ in abiotic marine carbonate cements (dashed and solid lines) result from oceanographic/climatic factors such as variation in the amount of isotopically depleted water stored in polar ice caps, deep oceanic circulation of polar water masses, change in temperature of surface tropical waters, etc. The $\delta^{18}O$ and $\delta^{13}C$ values shown on this diagram are best estimates of the primary isotopic compositions of marine carbonate cements determined by the convergence of $\delta^{18}O$ and $\delta^{13}C$ diagenetic trends to their heaviest values for each geological period. The inset displays the same data in the more familiar $\delta^{18}O$ vs δ^{13} crossplot.

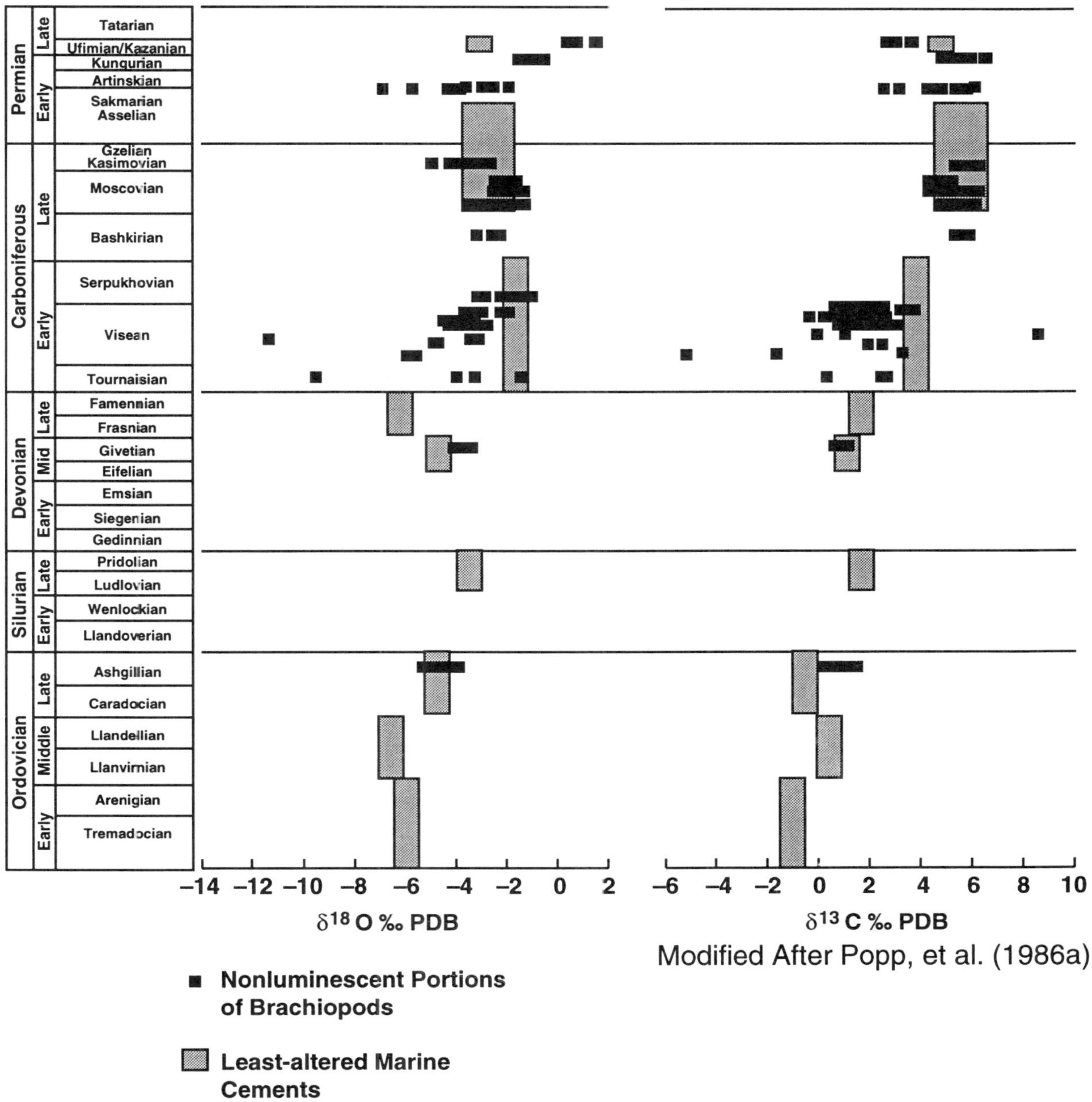

Figure 17. $\delta^{18}O$ and $\delta^{13}C$ data from portions of brachiopods screened for alteration by CL microscopy (solid squares; from Popp, et al., 1986a) are compared to estimates of "least-altered" $\delta^{18}O$ and $\delta^{13}C$ values for marine carbonate cements (shaded rectangles, from Lohmann and co-workers). Marine invertebrate and marine cement data sets agree in a general way, but can be seen to diverge in several intervals. Reasons for divergence are discussed in Appendix II.

Our approach in this manual has been to collect data from many sources and establish *ranges* of most-enriched $\delta^{18}O$ values for marine carbonate material for each geological period.[3] Carbon and oxygen isotope ranges for marine carbonate material spanning the entire Phanerozoic were gathered from more than 45 papers and are listed in Table II-1 in Appendix II. The most-enriched $\delta^{18}O$ values from each data set were used to establish a range of "least-altered" $\delta^{18}O$ values, which represents the marine carbonate baseline for each geological period. The range for each geological period reflects the sum of geographical and temporal variations in $\delta^{18}O$ within that period. The results are summarized in Figs. II-1 through II-14 . The method used to construct these figures is described in Appendix II.

Figure 18 presents the relationship that allows the marine carbonate isotopic baseline to be used for interpreting dolomite isotope data. The data used to construct Fig. 18 are listed in Tables II-1 and II-2 in Appendix II. Holocene and Plio-Pleistocene early supratidal, sabkha, marine, and marine-meteoric mixing zone dolomites, which formed at or near the earth's surface from marine brines, seawater, and mixed marine-meteoric fluids at low temperatures, have $\delta^{18}O$ compositions that are similar to or more ^{18}O-enriched than brachiopod, mollusc, and foraminiferal shell material and marine carbonate cements precipitated in equilibrium with Holocene seawater. This relationship between calcite and aragonite marine carbonate constituents and early dolomite is observed worldwide, even though early dolomitizing fluids are more isotopically variable than seawater and a fractionation of several per mil exists between calcite and dolomite at earth surface temperatures. Early dolomites which formed in other geologic periods should show similar relationships to marine invertebrates and marine carbonate cements formed during those periods. Late burial dolomites, which formed at higher temperatures, should be ^{18}O-depleted compared to their respective marine carbonate baselines.

Figures 19A and 19B present examples of the application of this technique to dolomites from the Permian of the Arabian Gulf and the Ordovician of the Michigan Basin. Shown on these figures are the $\delta^{18}O$ ranges of "least-altered" marine invertebrates and marine carbonate cements that define the marine carbonate baselines for the Permian and Ordovician periods. Note that the Arabian Gulf dolomites have $\delta^{18}O$ values that are coincident with the Permian marine carbonate field, indicating a low temperature origin, while the Michigan Basin dolomites are significantly lighter than the Ordovician marine carbonate field, indicating a high temperature origin. Plots of this type can be generated for dolomite of any age back to the Cambrian using Figs. II-1 to II-14 in Appendix II.

Recent dolomites exhibit variable stoichiometry, structural disordering, and sector zoning, which are not usually retained in older, recrystallized and stabilized dolomites. Recrystallization and stabilization may result in a shift in $\delta^{18}O$ to more depleted values. Thus, a sabkha dolomite might lose its low temperature oxygen isotope signature during burial. Fortunately, as we shall see in the next section, complete re-equilibration is rare. However, it is useful to have another approach for interpreting dolomite origin.

---

[3] Carbon isotope fractionation in carbonates does not have a significant temperature dependence and the $\delta^{13}C$ composition of low temperature dolomite varies widely, due to incorporation of carbon from different sources in the dolomite lattice. Thus, the carbon isotopic composition of dolomite does not provide particularly useful information for discriminating low temperature and high temperature dolomites. However, in carbonate diagenesis studies it is conventional to plot $\delta^{18}O$ versus $\delta^{13}C$. Therefore, in this discussion, data are displayed in $\delta^{18}O$ vs $\delta^{13}C$ crossplots, although only the $\delta^{18}O$ data are considered significant in our interpretations.

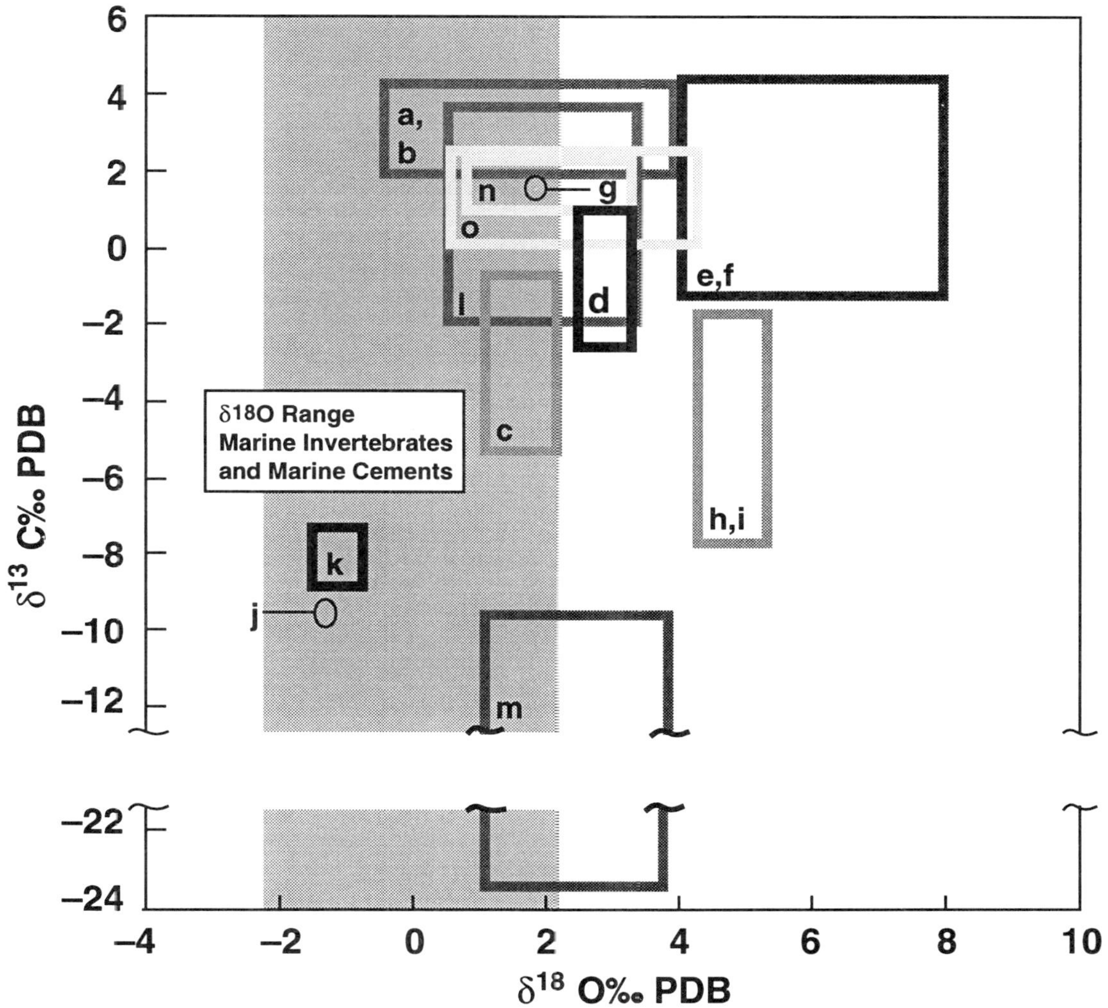

Figure 18. Carbon and oxygen isotopic compositions of Holocene and Plio-Pleistocene "early" sabkha, supratidal, subtidal marine, and marine-meteoric mixing zone dolomite (boxes a through n) compared to $\delta^{18}O$ range of Holocene brachiopods, molluscs, foraminifera, and marine carbonate cements, which are composed of low-Mg calcite, high-Mg calcite, and aragonite (shaded area). Early dolomite has $\delta^{18}O$ compositions similar to or isotopically heavier than marine carbonate material. "Late" high temperature dolomite formed during burial should have more negative $\delta^{18}O$ values and should plot to the left of the shaded area. Sources of data for Holocene marine carbonate material (shaded area) are listed in Appendix II, Table II-1. Sources for "early" dolomite (boxes a through n) are listed in Appendix II, Table II-2.

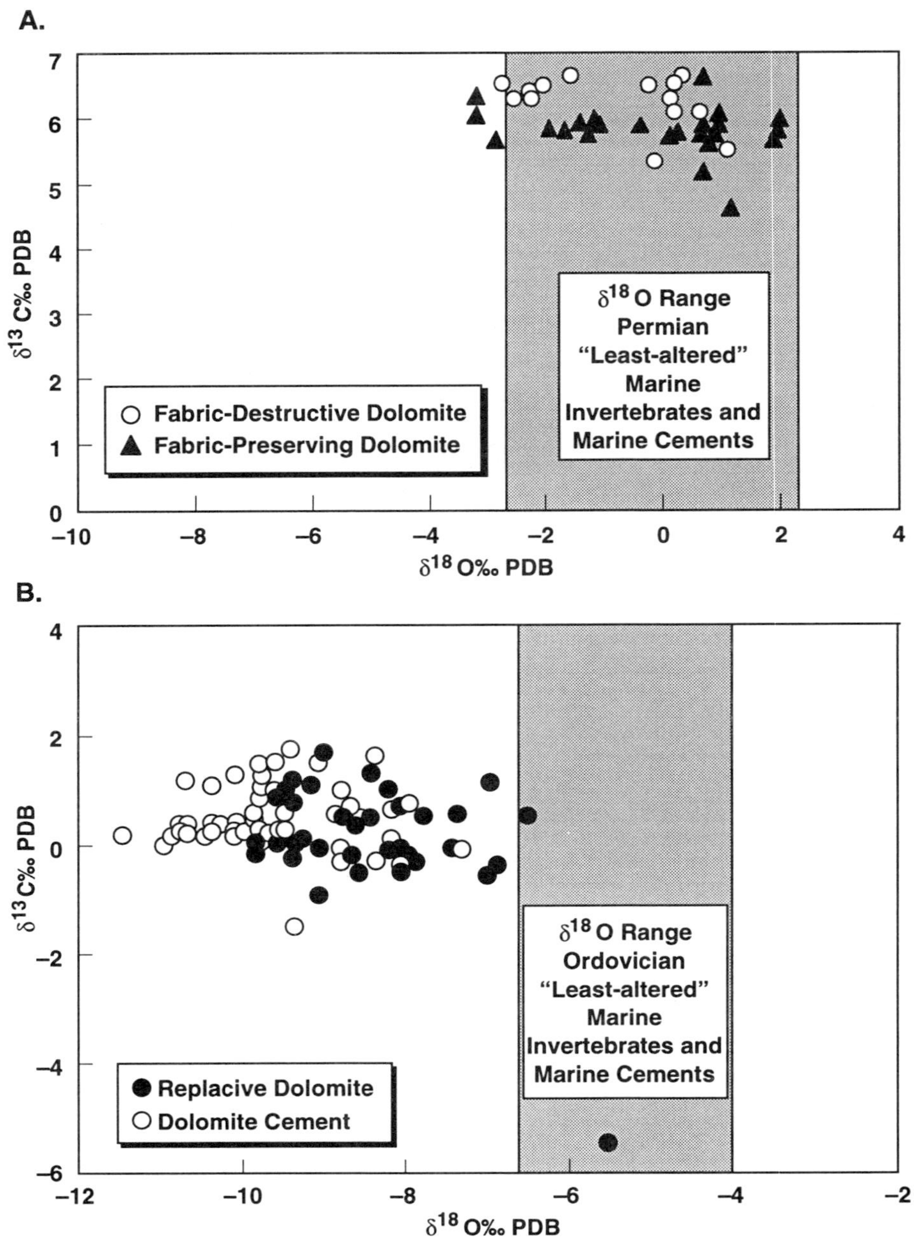

Figure 19. (**A**) Carbon and oxygen isotopic compositions of fabric-preserving and fabric-destructive replacive dolomites from the Permian of the Arabian Gulf are compared to the $\delta^{18}O$ range of "least altered" (i.e., isotopically heaviest) Permian marine invertebrates and marine carbonate cements. Most dolomite falls within the range established for Permian marine carbonate material, indicating low temperature origin for the dolomites. Several values fall to left of the Permian marine carbonate field. Since these samples are texturally identical to the other dolomite, their $\delta^{18}O$ compositions have probably been shifted to lighter values by partial recrystallization at higher temperatures during burial. (**B**) Carbon and oxygen isotopic compositions of replacive dolomite and dolomite cement from the Ordovician of the Michigan Basin are compared to $\delta^{18}O$ range for "least altered" (i.e., isotopically heaviest) Ordovician marine invertebrates and marine carbonate cements. Both replacive dolomite and dolomite cement are significantly depleted in ^{18}O compared to Ordovician marine carbonate material, indicating precipitation or recrystallization occurred at elevated temperatures. Similarity in isotopic composition between replacive dolomite and dolomite cement suggests they may have formed from the same fluids at similar temperatures.

APPROACH #2. COMPARISON TO OTHER DOLOMITES

Another test which can be used to determine the origin of an unknown dolomite is the comparison of its isotopic composition to isotope ranges for "low temperature" and "high temperature" dolomites whose origins have been previously established by independent means. Independent lines of evidence may include petrographic characteristics (such as undulatory extinction), stratigraphic relationships (such as concordance with or cross-cutting of depositional facies), and other geochemical parameters (e.g., trace and minor elements, Sr isotopes, and fluid inclusions).

Early, low temperature dolomite is defined as dolomite that formed in supratidal, sabkha, reflux, marine, and marine-meteoric mixing zone environments at or near the earth's surface. Carbon and oxygen isotope ranges for early low temperature dolomite from a variety of environments, localities, and ages are presented in Appendix III, Table III-1. Although the evidence used to establish dolomitizing environment was sometimes subjective, the waters responsible for dolomitization ranged in composition from brackish to hypersaline, and some of the early dolomite may have partially recrystallized during later burial, all but 4 of the 64 examples of low temperature dolomite listed in Appendix III, Table III-1 have $\delta^{18}O$ values more positive than about -6.5‰ PDB (Figs. 20 and III-1).

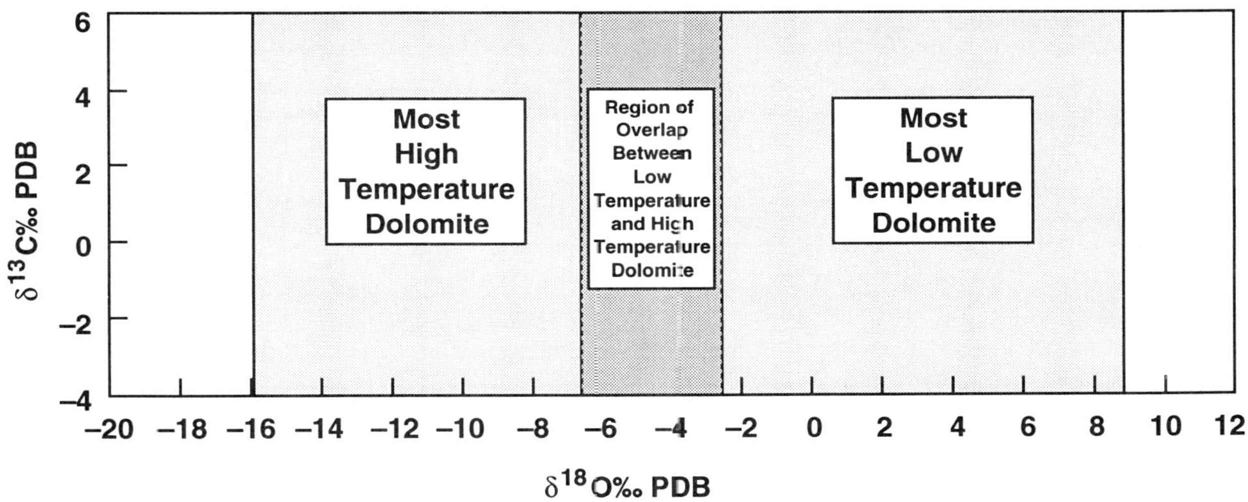

Figure 20. Graphical summary of data tabulated in Appendix III, Tables III-1 and III-2. Low temperature sabkha, supratidal, reflux, marine, and marine-meteoric mixing zone dolomite from formations of a variety of ages that span most of the Phanerozoic have $\delta^{18}O$ values that range from -6.5 to +9.0‰ PDB (Appendix III, Fig. III-1). High temperature burial and gangue dolomite and dolomite cement from formations of various ages have $\delta^{18}O$ values that range from -16 to -2.5‰ PDB (Appendix III, Fig. III-2). When plotted on a $\delta^{13}C$ vs $\delta^{18}O$ crossplot, the two data fields overlap between $\delta^{18}O$ values of -6.5‰ and -2.5‰ PDB. However, a low temperature dolomite field, with $\delta^{18}O$ values more positive than -2.5‰ PDB and a high temperature dolomite field with $\delta^{18}O$ values more negative than -6.5‰ PDB are clearly defined.

Late, high temperature dolomites form during burial, typically as replacement phases; as cement that fills fractures, vugs, and intergranular pore space; and as hydrothermal gangue dolomite associated with ore deposits. Late dolomites commonly contain two-phase fluid inclusions with homogenization temperatures of >50°C. Many have curved or "saddle-shaped" crystal faces and display undulatory extinction—both features are generally associated with precipitation at elevated temperatures (Radke and Mathis, 1980). All but 1 of the 54 examples of high temperature dolomite listed in Appendix III, Table III-2 have $\delta^{18}O$ values which are more negative than about -2.5‰ PDB (Figs. 20 and III-2).

Thus, one can conclude that dolomite which is more positive than -2.5‰ PDB most probably had a low temperature origin and dolomite which is more negative than -6.5‰ PDB most probably had a high temperature origin. However, a region of overlap occurs between -2.5 and -6.5‰. In this region, a dolomite could have had either origin (Fig. 20).

Because of the uncertainties involved in determining the original environments of dolomites used to establish the fields and the partial recrystallization of some low temperature dolomites during burial, the field boundaries should not be taken as absolute, but merely as useful guides for interpreting dolomite origin. Exceptions may be found, but the general positions of the field boundaries are well-supported by the data.

The Arabian Gulf and Michigan Basin data shown in Figs. 19A and 19B, which compare these dolomites to marine carbonate baselines, have been replotted on high temperature/low temperature dolomite plots in Figs. 21A and 21B. Comparison of Figs. 19 and 21 shows that both approaches yield the same interpretation.

Figures 22A and 22B show how the two approaches can be used together to provide a cross-check and additional interpretative power. Fig. 22A compares isotope data for dolomite from a Cretaceous formation in the Gulf of Mexico area to the $d^{18}O$ range for Cretaceous "least-altered" marine invertebrates and marine carbonate cements. Examination of Fig. 22A alone might lead one to conclude that the dolomite in well nos. 1 and 2 had an early low temperature origin while the dolomite in well nos. 3, 4, and 5 had a late, high temperature burial origin. Figure 22B, however, suggests that only the dolomite in well nos. 4 and 5 are conclusively high temperature. The $d^{18}O$ values for the dolomite in well no. 3 fall in the region of overlap between low temperature and high temperature dolomite and could be either burial dolomite or low temperature dolomite that partially recrystallized during burial.

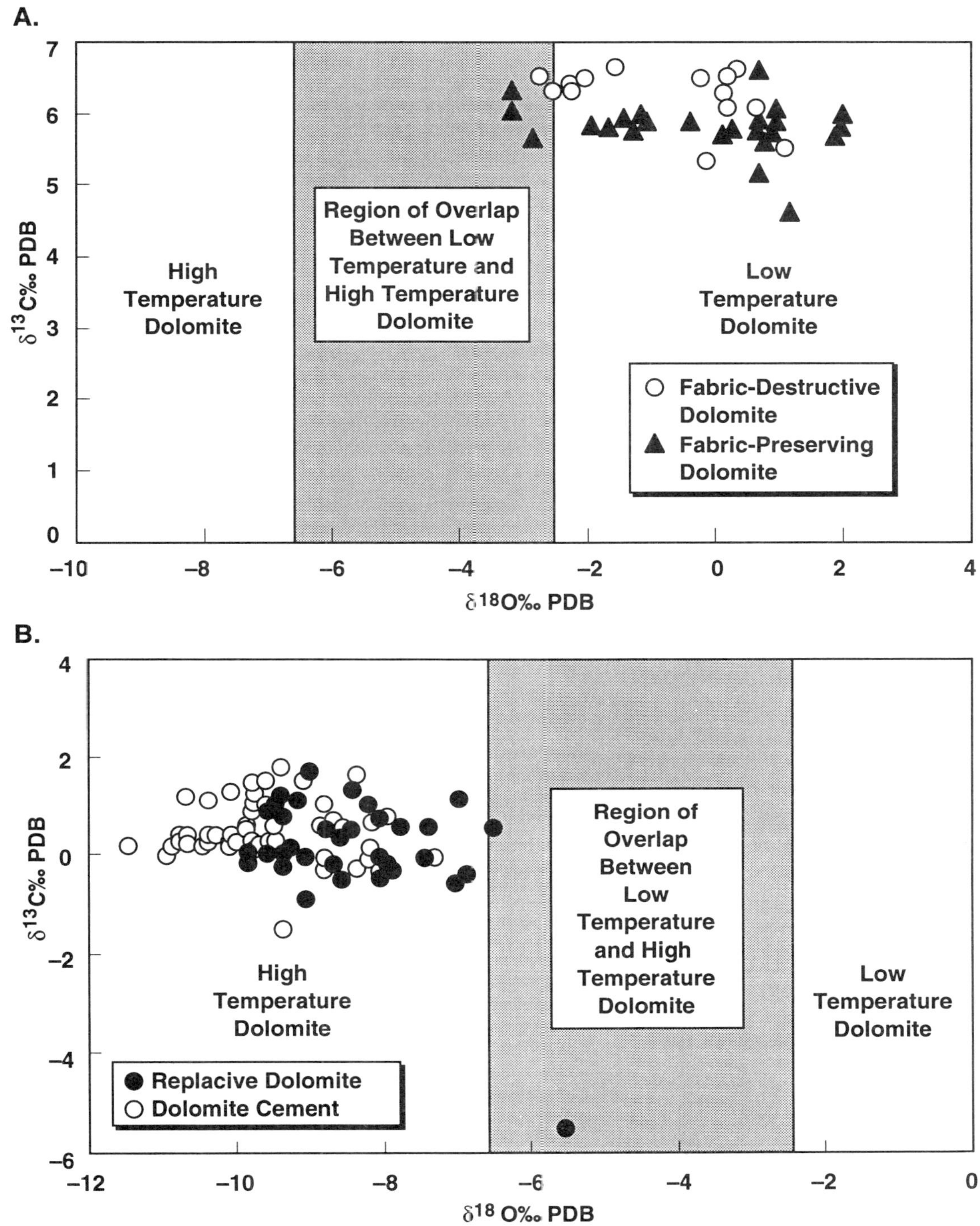

Figure 21. (A) Same carbon and oxygen isotope data shown in Fig. 19A for dolomite from Permian of the Arabian Gulf is replotted here on high temperature/low temperature dolomite plot. All but two values fall in low temperature dolomite field, which is consistent with interpretation arrived at in Fig. 19A using marine carbonate baseline approach. (B) Same carbon and oxygen isotope data shown in Fig. 19B for dolomite from Ordovician of the Michigan Basin is replotted here on high temperature/low temperature dolomite plot. All but two values fall in high temperature dolomite field, which is consistent with interpretation arrived at in Fig. 19B.

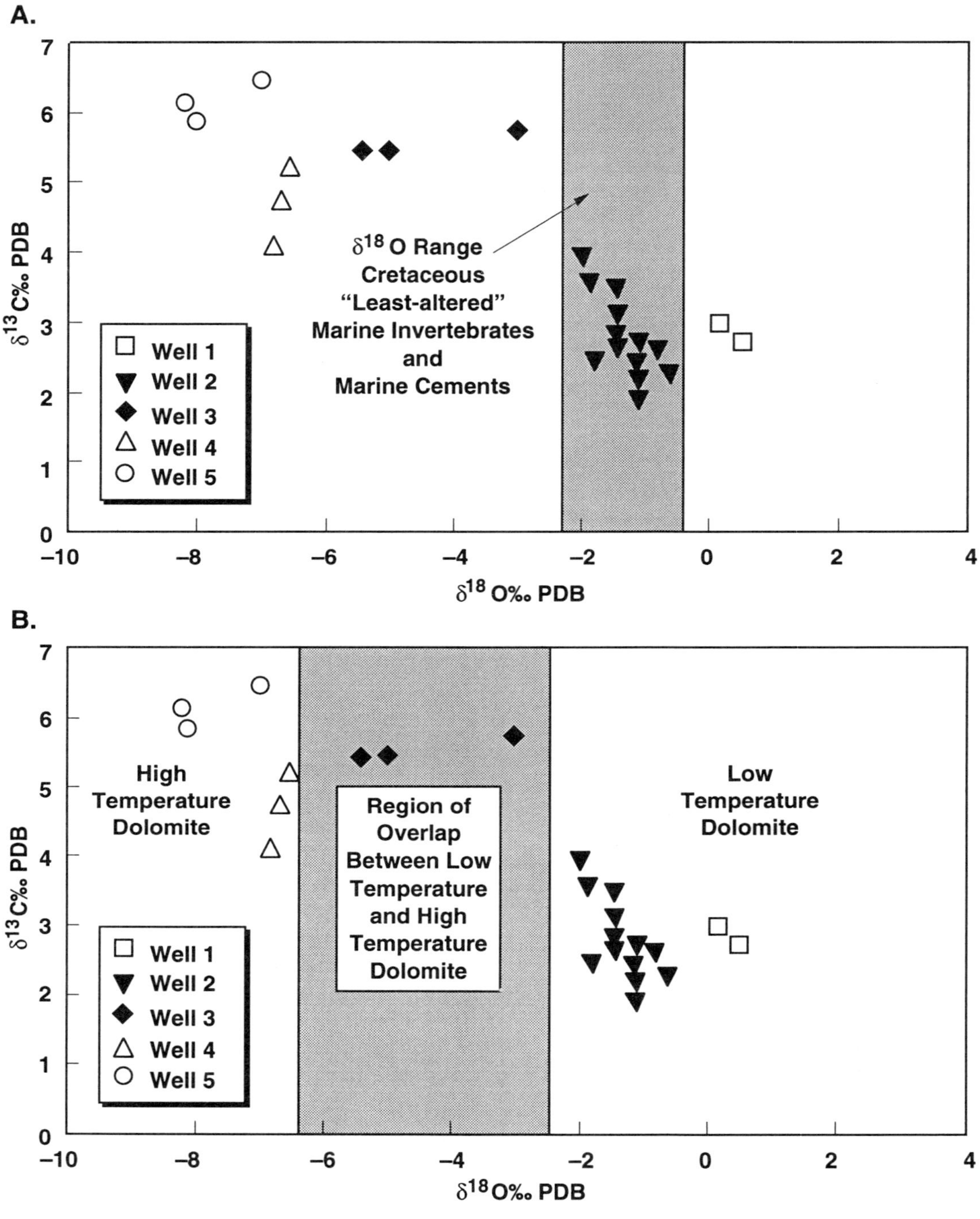

Figure 22. (A) Isotopic compositions of dolomites from 5 wells which penetrate a Cretaceous formation in the Gulf of Mexico area are compared to d[18]O range of "least-altered" (i.e., isotopically heaviest) Cretaceous marine invertebrates and marine carbonate cements. Dolomite in Wells 1 and 2 appears to have had early, low temperature origin, whereas dolomite in Wells 3, 4, and 5 appears to have had higher temperature origin. (B) Isotopic compositions of same dolomites as in Fig. 22A are shown here on a high temperature/low temperature dolomite plot. This plot reveals that dolomite in Wells 1 and 2 had a low temperature origin whereas dolomite in Wells 4 and 5 had a high temperature origin. Dolomite in Well 3 falls in region of overlap and could be either high temperature burial dolomite or low temperature dolomite that partially recrystallized during burial.

CALCULATING ISOTOPE TEMPERATURE

In the rare case where the isotopic compositions of both the dolomite and the dolomitizing fluid are known, dolomitization temperature can be calculated. To understand better how this is done, consider as an example a dolomite with a δ value of -6.5‰ PDB and seawater (SMOW) as the parent fluid. At what temperature did this dolomite crystallize?

The solution is obtained by first converting the dolomite $\delta^{18}O$ value from the PDB scale to the SMOW scale, using Eq. (iv) in Appendix I:

$\delta^{18}O$ (SMOW) = [1.03086 * (-6.5)] + 30.86 = +24.1;
Using Eq. 1, the temperature is then calculated:
(24.1) - (0) = $[3.2 * 10^6 * T(°K)^{-2}] - 1.5$;
and, rearranging:
$T(°K) = [(3.2 * 10^6)/(24.16 + 1.5)]^{1/2}$ = 353°K, or 80°C.

An alternative graphical method to obtain temperature from $\delta^{18}O_{dol}$ is given in Figure 23.

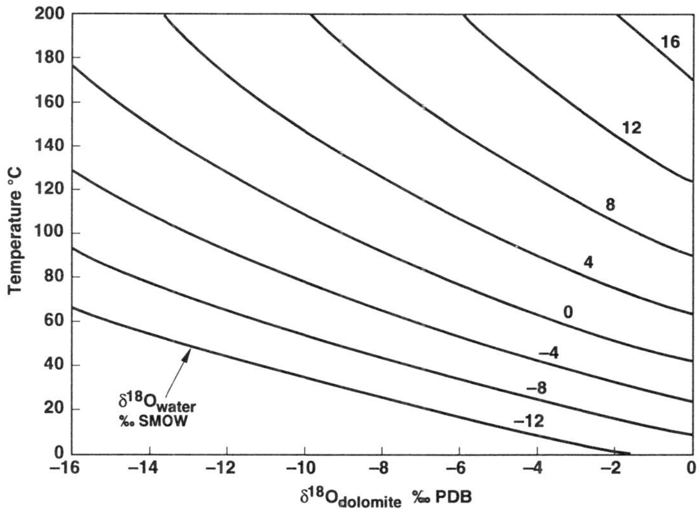

Figure 23. Plotted are the curves for various values of $\delta^{18}O_{water}$ as a function of both $\delta^{18}O_{dol}$ and temperature (T) of dolomite precipitation, according to the relationship (Friedman and O'Neil, 1977):

$$\delta^{18}O_{dol} - \delta^{18}O_{water} = [3.2 * 10^6 T(°K)^{-2}] - 1.5$$

These curves can be used to estimate dolomitization temperature if $\delta^{18}O_{dol}$ is measured and if $\delta^{18}O_{water}$ is reasonably-well constrained. Alternatively, if temperature is known, for example from fluid inclusion Th values, then $\delta^{18}O_{water}$ can be estimated (see text).

The oxygen isotopic compositions of waters of interest to us span a range of at least 10‰. Because we have no knowledge of an ancient water's $\delta^{18}O$, it follows that temperatures calculated from $\delta^{18}O_{dol}$ are limited by our educated attempts to constrain the value of $\delta^{18}O_{water}$. By way of illustration of the magnitude of the error involved, consider a water whose composition is estimated to within 10‰ SMOW, and from which a dolomite with a $\delta^{18}O_{dol}$ of -6.5‰ PDB has formed. Temperatures calculated from Eq. 1 for this dolomite span a range of 71°C:

$\delta^{18}O_{water}$ ‰ SMOW (ASSUMED)	Temperature (°C) (CALCULATED)
+5	121
0 (=SMOW)	80
-5	50

In general, temperatures calculated from $\delta^{18}O_{dol}$ are changed by approximately 5-8°C for each 1‰ change in estimated $\delta^{18}O_{water}$.

In summary, the best methods for discriminating low and high temperature dolomite are empirical: (1) by comparison to unaltered marine carbonate material, and (2) by comparison to well-studied dolomites. Ideally, we would like to calculate the temperature of dolomitization accurately, but this is valid only if $\delta^{18}O_{water}$ can be constrained to within a few per mil.

CARBON ISOTOPES IN DOLOMITE

Carbon isotope ratios are measured simultaneously with oxygen isotope ratios on mass spectrometers. The carbon isotopic composition of dolomite reflects the ratio of inorganic carbon derived from pre-existing limestone and dolomite to organic carbon derived from the microbial and thermal breakdown of organic matter. $\delta^{13}C$ data on marine invertebrates and marine carbonate cements for each geologic period in the Phanerozoic are compiled in Table II-1, Appendix II. The most-enriched $\delta^{13}C$ values of unaltered carbonates can serve as baselines against which a dolomite can be compared and screened for organic-carbon contributions. The marine cement data base discussed in Lohmann and Walker (1989) and summarized in Fig. 16 can be used for this purpose. $\delta^{13}C_{dol}$ values that deviate by more than a few per mil from a marine baseline that is contemporaneous with time of dolomitization signify organic contributions.

The $\delta^{13}C$ compositions of some important carbon reservoirs and their isotope ranges are shown in Figure 24. Note that a chemical reaction must convert the organic matter to CO_2 or a carbonate species such as HCO_3^- before organic carbon can be incorporated into the dolomite lattice. Specific organic reactions and the approximate range of $\delta^{13}C$ compositions of dolomites precipitated in conjunction with these reactions are listed in the following table:

	Process	**Reaction**[4]	**$\delta^{13}C_{dol}$**
i.	Oxidation	$CH_2O + O_2 \rightarrow CO_2 + H_2O$	-5 to -15
ii.	SO_4 reduction	$2CH_2O + SO_4^{-2} \rightarrow 2HCO_3^- + H_2S$	-5 to -15
iii.	Fermentation	$2CH_2O \rightarrow CH_4 + CO_2$	+5 to +20
iv.	Methane oxidation	$CH_4 + 2O_2 \rightarrow CO_2 + 2H_2$	<-25
	or oil degradation	$4CH_2 + 3SO_4^{-2} \rightarrow 4HCO_3^- + H_2S + 2HS^-$	-5 to -15
	or "TSR"[5]:	$CH_4 + SO_4^{-2} \rightarrow HCO_3^- + HS^- + H_2O$	<-25?
v.	Decarboxylation	kerog' $\rightarrow CO_2$ + kerog''	-5 to -15

The significance of each of these pathways is briefly discussed next.

Oxidation involves the generation of CO_2 or bicarbonate (HCO_3^-) from the bacterial degradation of organic matter within a soil profile, or beneath the sediment-water interface in well-oxygenated pore fluid. The groundwater or pore water CO_2 in reaction (i) may influence the isotopic composition of dolomite that formed in shallow marine or mixed, marine-meteoric water. A carbonate mineral which incorporates carbon from 100% marine-organic or soil-organic sources will have a $\delta^{13}C$ composition close to -15‰ PDB, and if the carbonate mineral incorporates carbon from a 50:50 mixture of organic- and host-rock-sourced carbon, then its $\delta^{13}C$ composition will be closer to -5‰ PDB. If the pore fluid environment was one in which the supply of organic carbon was low, then the isotopic composition of dissolved CO_2 and diagenetic carbonate will register only inorganic $\delta^{13}C$ values (this is the usual case).

---

[4] Organic matter = "CH_2O".

[5] "Thermochemical Sulfate Reduction"

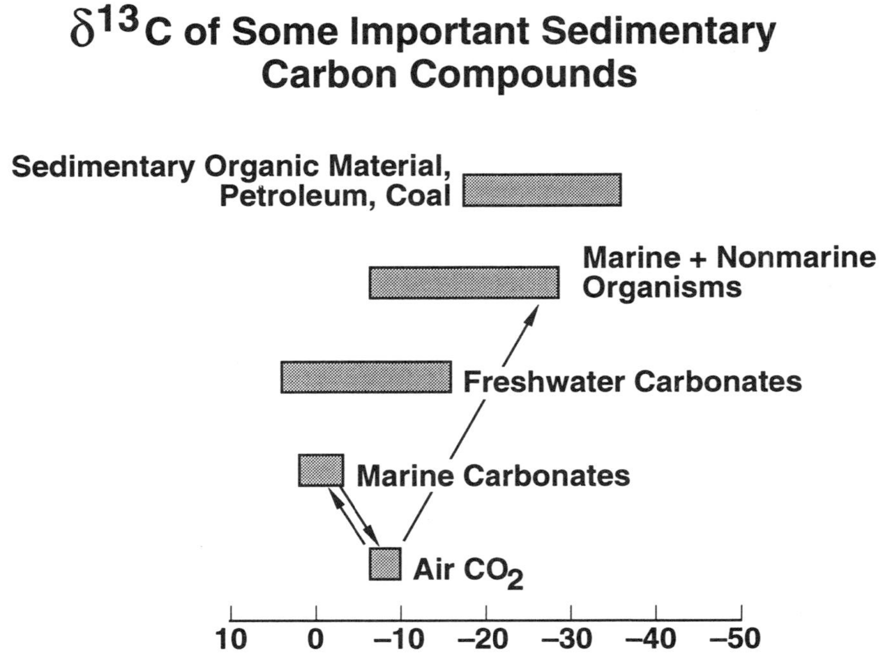

Modified after Hoefs (1980)

Figure 24. $\delta^{13}C$ composition of some common carbon-bearing, sedimentary compounds. The ultimate source of carbon in submarine cements, invertebrate shells and soft-parts, aquatic algae, marine kerogen and oils is atmospheric CO_2. Carbon isotopes are fractionated such that ^{12}C is concentrated in organic material and the products of its degradation (microbial methane, kerogen, oil), and ^{13}C is concentrated in inorganic material such as shells, cements and limestone.

Bacterial sulfate reduction (ii) and fermentation (iii) occurring in anoxic pore fluids below the marine sediment-water interface are important in sediments rich in volatile organic matter, and contribute to dolomite concretion formation. These processes are important where the supply of molecular oxygen is poor due to bacterial consumption of organic matter and oxygen. The maximum depth below the sediment-water interface at which sulfate reduction can occur depends on the sediment TOC content and the rate of supply of seawater-sulfate to the shallow sediment column. In general, seawater sulfate is exhausted within a few meters of the sediment-water interface in high TOC sediments, but may persist to 20 m depth if the sediment TOC is less than 0.5 wt%. A byproduct of sulfate reduction is the generation of H_2S (Trudinger, et al., 1985) and precipitation of pyrite, which is common in most marine carbonates.

Below the diffusion limit for seawater-sulfate, methane-producing bacteria predominate. Reactions producing methane have very distinctive isotope signals. Bacterial or microbial methane commonly has a $\delta^{13}C$ composition of <-60‰ PDB, leaving CO_2 co-produced with the methane from the same hydrocarbon molecule with a distinctively "heavy" $\delta^{13}C$ value. $\delta^{13}C$ of any dolomite formed from this CO_2 is generally between +5 and +20‰ PDB. Microbial methane usually forms shallower than 1 km depth ("fermentation" in Fig. 25); thus, $\delta^{13}C$ values in dolomite more than 5‰ heavier than the marine baseline indicate shallow subsurface dolomitization. The "early" dolomite concretions in the Miocene Monterey Formation of California fall into this category.

CO_2 from methane oxidation (iv) is also isotopically distinctive because it inherits methane's light $\delta^{13}C$ composition. The oxidation of methane using SO_4 at high temperatures (>80°C; Krouse, et al., 1988; Sassen, 1988) is called "thermochemical sulfate reduction", or TSR. TSR can operate on any organic substrate, including oil, and results in (1) the generation of H_2S gas, (2) the removal of normal paraffins in oil, and (3) the simultaneous precipitation of pyrite, native sulfur, and late dolomite cement. The carbon isotope fractionation for TSR is unknown, probably because TSR is not a simple, single-step process. Dolomite produced by TSR may have $\delta^{13}C$ values that range from -2 to -12‰ lighter than the contemporaneous marine $\delta^{13}C$ composition (Machel, 1987), although cases have been documented in which little or no isotope fractionation occurs (Heydari and Moore, 1989). TSR is an important mechanism for generating the high H_2S contents in formation fluids and oils which are observed in many basins (e.g., the Permian and Alberta basins).

Thermal decarboxylation (v) is a poorly understood process. The thermal breakdown of kerogen releases CO_2. The process begins prior to liquid hydrocarbon generation and may persist into the oil window, although the various stages of decarboxylation have not been well-documented. Decarboxylation is often complete at maturation levels corresponding to a vitrinite reflectance of 0.50% in the Texas Gulf Coast.

CO_2 and methane may be contributed to pore water even deeper in the section from hydrocarbon cracking reactions, and from thermal decarboxylation and sulfate reduction of organic acids (above 90°C; Kharaka, et al., 1983).

The carbon from kerogen and organic acids which matured or decomposed deep within a basin may ultimately find its way into dolomite in shallower parts of the basin. This can happen when fluids ascend along faults and outgas CO_2 with decreasing depth and pressure. Loss of CO_2 causes dolomite precipitation in fractures and vugs. This process results in the ^{13}C depletion observed in many late dolomite cements.

Figure 25 presents some of these sources and their effects on the $\delta^{13}C$ of pore water CO_2 as a function of burial depth. The CO_2 that results from these sources (oil biodegradation, fermentation, etc.) is fractionated during the transformation to HCO_3^- and, ultimately, calcite, as shown in Figure 26. The total fractionation is about 10‰ between CO_2 and solid carbonate in equilibrium with it, so that a dolomite will be isotopically heavier than the CO_2 in the porewater from which it precipitates (Fig. 26; Emrich, et al., 1970).

Carbon isotope systematics are so complex that linking $\delta^{13}C_{dol}$ to a single carbon source (other than methane, which is isotopically distinctive) is difficult. Most importantly, mingling of "organic" carbon with "inorganic" (marine) carbon dilutes the "organic" isotope signal and creates a hybrid

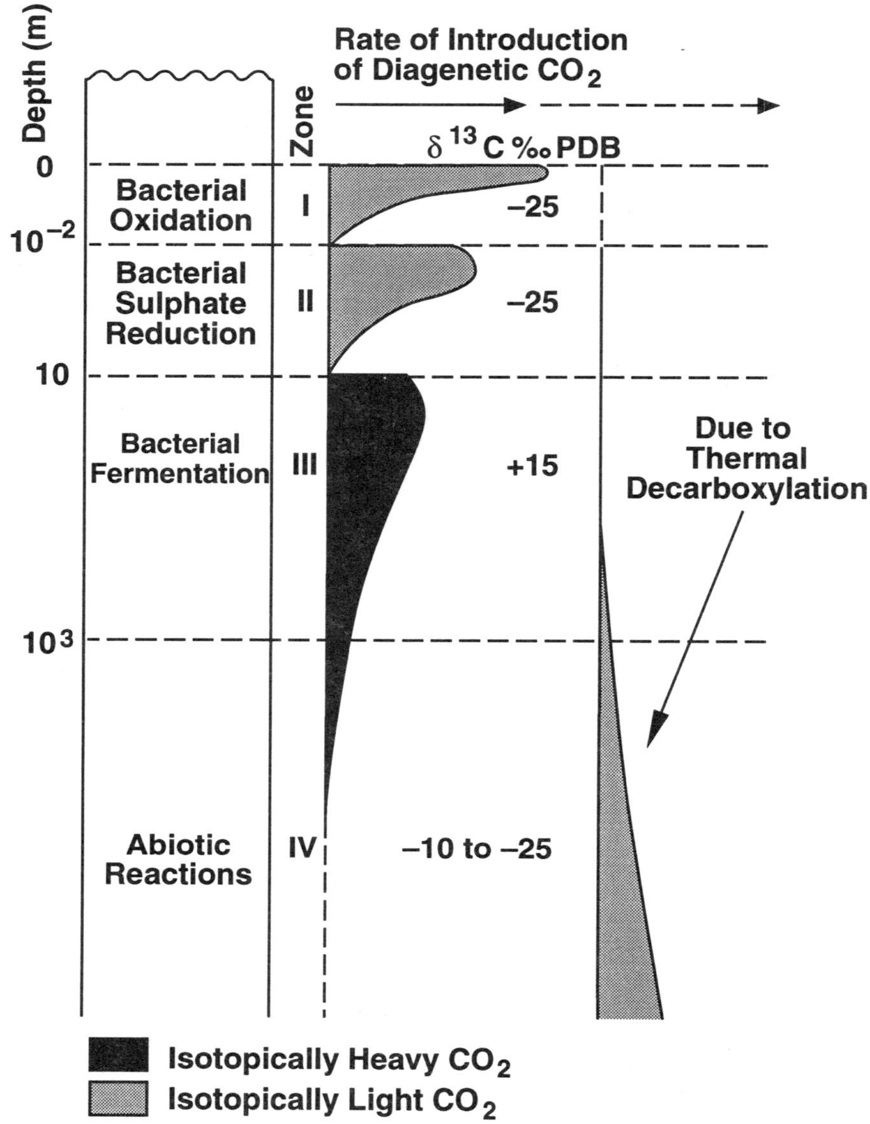

Modified after Irwin, et al. (1977)

Figure 25. Schematic illustration of the effects of microbial and thermal alteration of organic matter on the isotopic composition of porewater CO_2 with increasing depth of burial. This profile starts at the seawater-sediment interface. The depth to which bacterial oxidation may affect organic matter (Zone I) is limited by the rate of oxygen diffusion from the water column balanced against the rate of bacterial consumption of oxygen in the sediment. Oxygen supply is poorest in anoxic environments or in sediments where organic content is sufficiently high to continually consume oxygen. Oxygen supply is greatest in zones of freshwater recharge. The thickness of the sulfate-reduction zone (Zone II) is controlled by the rate of seawater sulfate diffusion from the water column balanced against sulfate consumption by sulfate-reducing bacteria. Bacterial fermentation or methanogenesis (Zone III) is limited to depths beyond the diffusion limit for sulfate and dissolved oxygen; therefore, there is usually a distinct boundary between Zones II and III. Zones II and III are commonly sites of carbonate concretion growth. Thermogenic processes dominate in Zone IV.

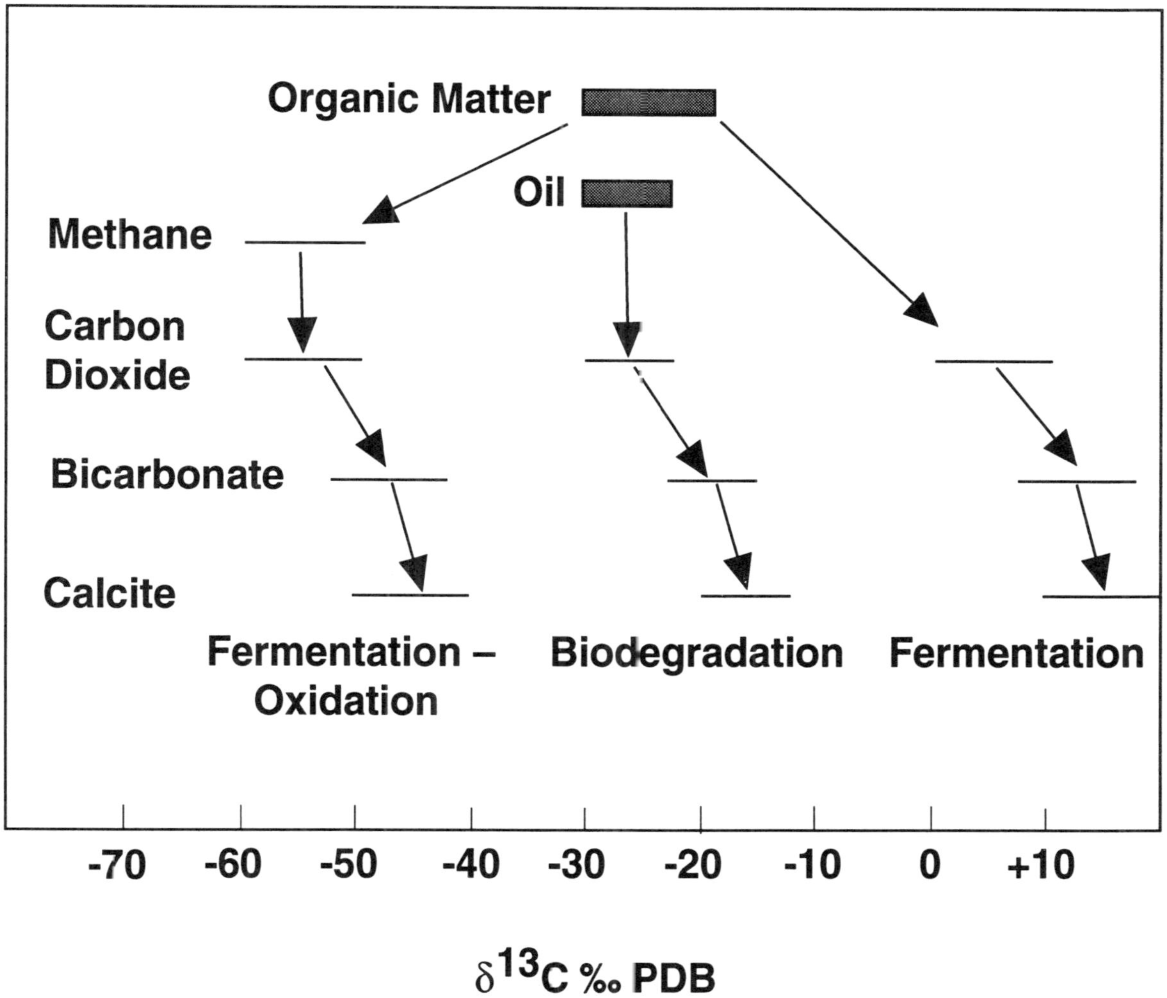

Modified after Wiggins, et al. (1993)

Figure 26. Schematic illustration of the route organically-sourced carbon takes before being precipitated in a carbonate mineral. Changes in the isotopic compositions of the carbon species are illustrated. The change from methane to carbon dioxide may also be accompanied by a bacterial fractionation (not shown; Barker and Fritz, 1981; James and Burns, 1984). The carbon isotope technique affords the ability to discriminate different sources of carbon and different reactions that incorporate the carbon into carbonate minerals.

carbon isotope signature that depends largely on the relative proportion of carbon derived from each source. Therefore, it is usually best merely to interpret the carbon isotopic composition of a dolomite in terms of presence or absence of organic carbon.

Figure 27 summarizes many of the factors controlling the carbon and oxygen isotopic compositions of carbonate rocks, including dolomites.

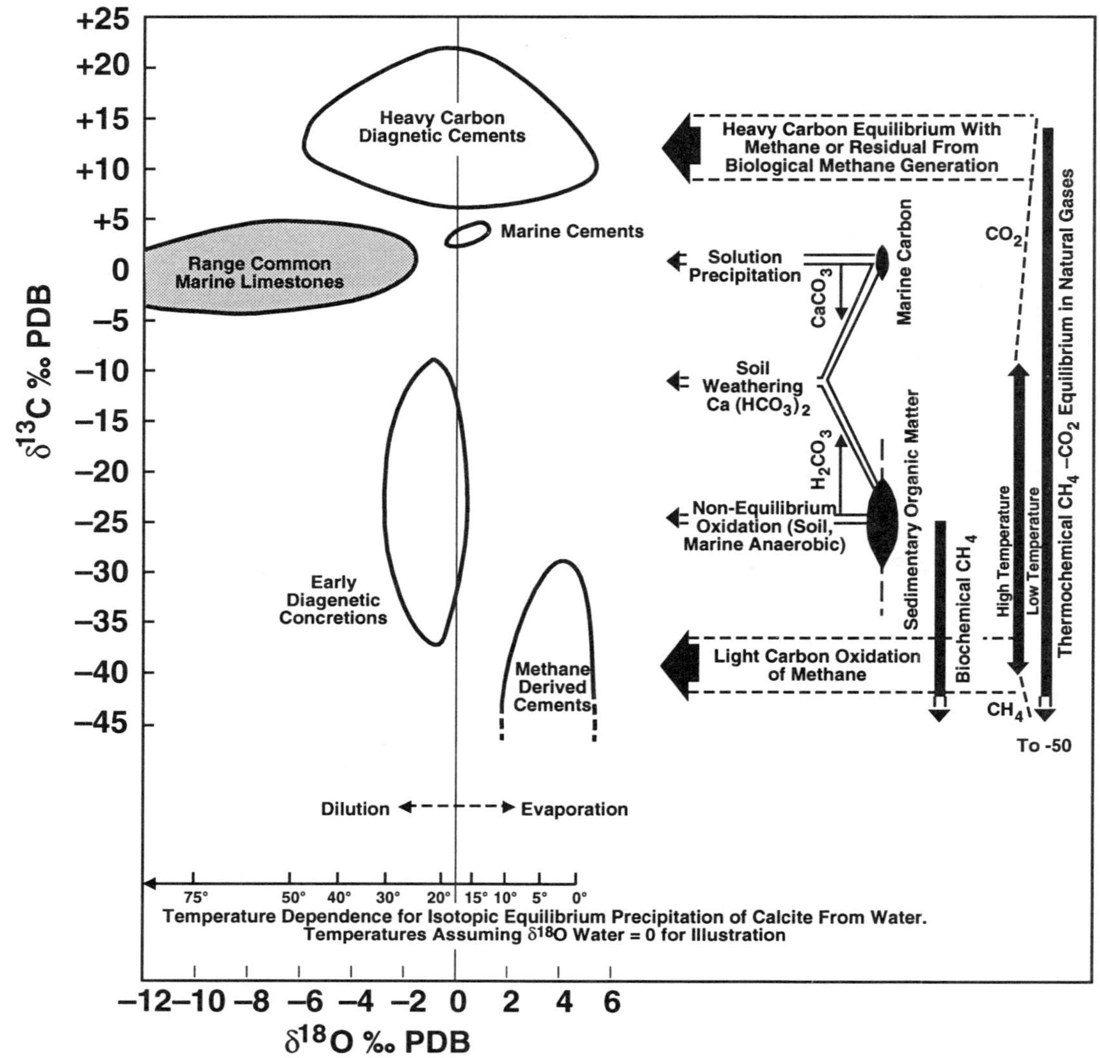

Modified after Hudson (1977)

Figure 27. Schematic summary of ranges of isotopic compositions in several carbonate systems. To the right along the vertical axis is a list of some of the important organic degradative processes discussed in the text. The horizontal axis (bottom part) shows the temperature dependance and salinity controls of the $\delta^{18}O$ composition of carbonate minerals. The shaded area, "range common marine limestones", is a generalized area in which values for carbonates of many different origins overlap. To more fully understand the origin of carbonates that fall within this range, one must collect additional types of data, such as fluid inclusion Th and Tm, and Sr isotope ratios.

FLUID INCLUSIONS

In an earlier chapter on formation waters, we outlined an approach for evaluating the role of present-day formation waters (brines) in dolomitization. In many cases, the dolomitizing fluid has long since vanished and has been replaced by one or more subsequent generations of fluid. The only direct record of the primary fluid in such cases is contained in fluid inclusions.

Fluid inclusions are fluid-filled cavities, usually 1-10 μm in diameter, in carbonate minerals (Roedder, 1981). Fluid inclusions form as pore fluid is trapped within lattice defects in actively growing crystals. They are possible only where crystal growth is within a pore or fracture, explaining why cements are better candidates for fluid inclusion studies than replacement dolomites.

Primary inclusions, those that represent the fluid that precipitated the dolomite and form as the crystal grows, can be trapped along a crystal growth zone or can be randomly distributed within a crystal (Roedder, 1981). Secondary inclusions, those which represent a fluid that did not precipitate the dolomite, and are incorporated into the crystal after it has grown, are trapped along microfractures that cut the external surface of the crystal. Primary inclusions are most easily identified along growth zones in crystals that fill fractures, vugs, or occur as overgrowths on sucrosic dolomite. Unzoned replacement dolomite commonly contains fluid inclusions as well, but, in the absence of growth textures, it is impossible to determine whether these inclusions are primary or secondary. Therefore, the most reliable fluid inclusion data are collected on dolomite cements.

The trapped fluid may consist of one or more of the following: (1) an aqueous liquid (typically a brine), (2) an aqueous vapor bubble, (3) a hydrocarbon liquid, or (4) a hydrocarbon gas (usually methane). Aqueous inclusions also may contain a halite daughter crystal, which precipitated after entrapment.

The distinction between "vapor" and "gas" is important. Vapor is a secondary phase which separates from the liquid in a liquid-filled inclusion, whereas gas is trapped as a primary phase with or without a co-existing liquid. This distinction is important when attempting to determine the pressure-temperature history of two-phase fluid inclusions. Fluid inclusions that contain a liquid phase accompanied by a vapor bubble were first trapped as a single liquid phase at an elevated temperature (>50°C) and pressure. In the process of cooling at the wellsite, or after uplift to outcrop, the liquid contracted and the inclusion separated into discrete liquid and vapor phases. Single-phase, liquid inclusions form at low temperatures (<50°C; Goldstein, 1986).

We seek three types of basic information in a study of fluid inclusions: (1) distinction of aqueous versus hydrocarbon liquids and gases (from fluorescence techniques), (2) fluid entrapment temperature (from the homogenization temperature of two-phase aqueous fluid inclusions), and (3) fluid salinity (from the melting temperature of aqueous fluid inclusions).

HYDROCARBON IDENTIFICATION

As every wellsite geologist and mudlogger knows, hydrocarbons fluoresce in cuttings which are exposed to long-wavelength UV (3200-2900 Å) excitation. Likewise, fluid inclusions that fluoresce under the same conditions contain hydrocarbon liquids (Burruss, 1981). Hydrocarbon composition, and therefore fluorescence, is controlled by hydrocarbon source type, thermal maturity, and aspects of migration and biological degradation (Castle, 1981).

The color of emitted fluorescence varies with API Gravity such that high-gravity oils and condensates have maximum emission at short wavelengths and medium- and low-gravity oils have maximum emission at longer wavelengths. A tabulation of API Gravity data from Riecker (1962) shows the following relationships:

Color Emitted	(Wavelength)	Approximate API Gravity
blue	(4440 Å)	43°
blue-green	(4830 Å)	39°
green-yellow	(5780 Å)	34°
red	(6300 Å)	15°

The intensity of fluorescence also varies with API Gravity, being greater for high-gravity oils (greater than 37° API) and decreasing for API Gravities below 25°.

Riecker (1962) analyzed several oils for aromatics, paraffins, naphthenes and resins. He noted that fluorescence is caused by many components in crude oil, but especially important are aromatics, which emit intense fluorescence in a wide range of colors (Fig. 28). Paraffins, which are common in higher API-Gravity oils and condensates in which aromatics have been cracked by thermal maturation, emit a blue background hue.

Because marine source rocks generate abundant aromatics, marine-sourced oils generally fluoresce brightly in various colors. Waxy (paraffinic) oils of non-marine sources, which may lack aromatics, emit a blue fluorescence. The use of fluorescence properties to infer source character is a very speculative venture, however, because oil maturation can induce a blue-shift by destroying aromatics. Other methods such as biomarkers have been used to determine the source type of crude oils.

Obviously, fluorescence has complex controlling factors, yet general relationships permit the use of fluorescence properties to help correlate hydrocarbon fluid inclusions with (1) produced oils, (2) other hydrocarbon fluid inclusions in the same sample, or (3) fluid inclusions in other samples from remote locations along a migration pathway. The correlation of hydrocarbon fluid inclusions to produced oils on the basis of shared fluorescence characteristics could be an additional way to identify an oil-migration pathway. Hydrocarbon fluorescence in fluid inclusions is an under-used technique which holds much promise for future studies. At present, however, we only use the

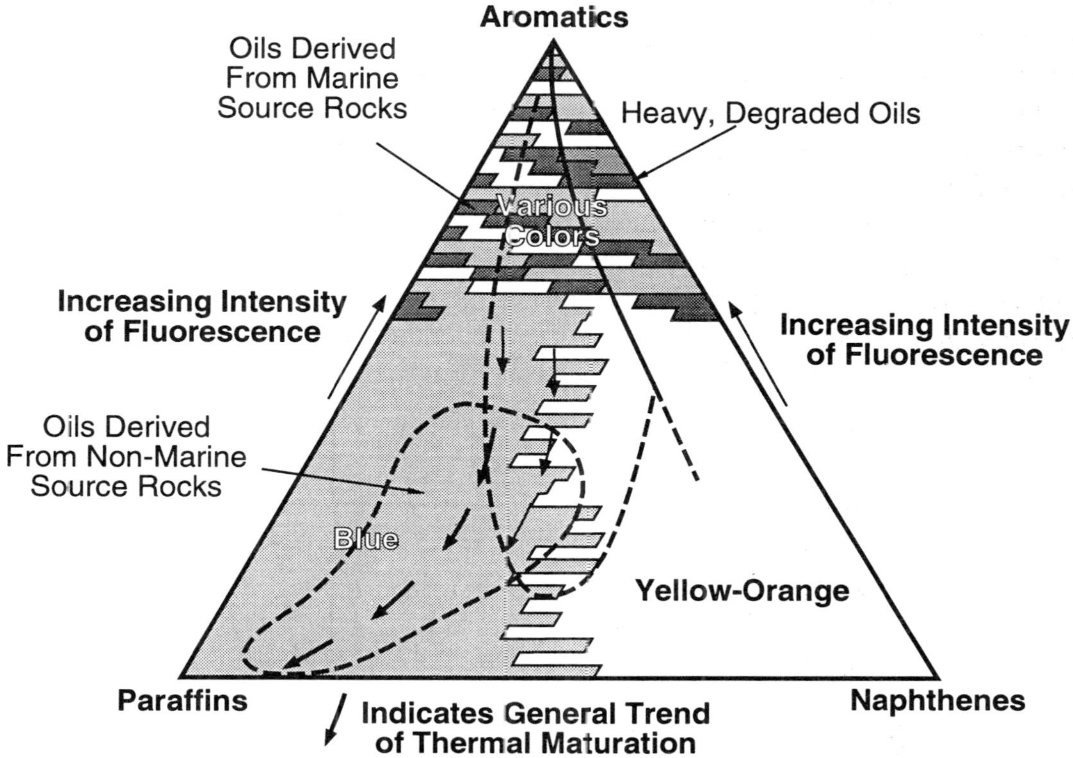

Castle (1981)

Figure 28. Ternary diagram of crude oil composition showing colors of fluorescence and direction of increasing fluorescence intensity. The effects of maturation override source effects in the determination of fluorescence color; therefore, caution is recommended in attempts to derive source type by this technique. The two most important applications of fluorescence in fluid inclusion analysis are differentiation of hydrocarbon from aqueous fluid inclusions and estimation of API Gravity. Biomarkers and other tools should be used to determine source type.

technique to identify the presence of hydrocarbons, estimate API Gravity, and determine the relationship between inclusions within the same sample (Burruss, 1991).

HOMOGENIZATION TEMPERATURE (Th)

The primary purpose of fluid inclusion studies is to establish the timing of dolomitization by matching entrapment temperature with a well's thermal history. In basin analysis, if one or several wells have been drilled, stratigraphic data, down-hole temperatures and heat flow history can be entered into a basin analysis program. Maturation profiles of vitrinite reflectance, spore coloration (TAI), apatite fission tracks, Rock Eval (Tmax) and porphyrin maturity parameters are often used in basin analysis studies to calibrate a subsidence profile to thermal history. An iterative process is employed in which thermal history is varied until a match to observed maturation indices is achieved. When this is completed, fluid inclusion entrapment temperature can be plotted on a thermal history curve to date dolomitization.

The liquid-vapor homogenization temperature (Th), the temperature at which a two-phase inclusion (liquid and its vapor) becomes one phase, provides an estimate of the minimum temperature of crystallization for the host crystal. The meaning of Th is appreciated by first considering a hypothetical inclusion history from time of entrapment to time of measurement in the laboratory. The density and volume of a fluid inclusion that is trapped as a single liquid phase at depth remains unchanged if the inclusion cavity does not stretch or leak (Crawford, 1981). The pressure and temperature conditions at entrapment can be represented on a phase diagram such as Figure 29. In this example, an aqueous fluid is trapped at the P and T conditions represented as Point A. Because the fluid has a density of 0.90 g/cc, Point A falls on the line of equal density (called an "isochore") corresponding to 0.90 g/cc. The 0.90 isochore is simply the locus of temperatures and pressures at which a fluid can maintain that density at constant volume.

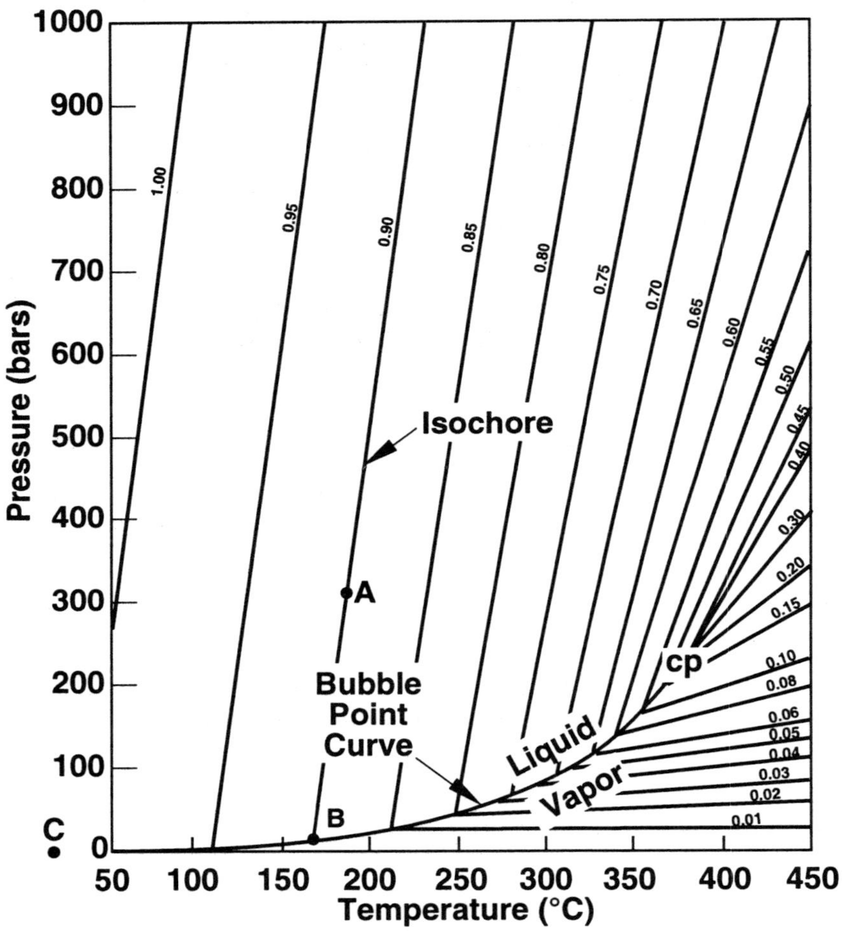

Crawford (1981)

Figure 29. On this phase diagram for pure water, let Point A represent the P,T conditions of a fluid inclusion's entrapment. During core retrieval, the pressure and temperature in the fluid inclusion decrease, but the fluid inclusion density remains unchanged. P and T decrease along the 0.90 g/cc isochore, which is a line of equal density. At Point B on the liquid-vapor curve ("bubble point curve"), liquid and vapor are both stable, and the liquid has contracted to the point where a vapor bubble nucleates on the wall of the inclusion. Cooling to room temperature causes the bubble to grow, and the inclusion ends up at Point C. Heating on a microscope stage results in re-homogenization of the vapor and liquid phases at Point B, which is known as the homogenization temperature, or Th. CP is the critical point for water.

When the core sample containing this fluid inclusion is recovered from the borehole, the fluid travels along the isochore through successively lower temperatures and pressures until it reaches the bubble point curve at Point B. At this point the fluid nucleates a vapor bubble. As the sample cools further at the wellsite, the liquid contracts and the bubble grows larger as the inclusion "travels down" the bubble point curve toward Point C.

The process is reversed in the laboratory by heating the inclusion to Point B, the point at which the bubble disappears and the fluid becomes a single phase. This point is the homogenization temperature (Th), and is regarded as the *minimum* temperature of entrapment— an approximation of the true temperature of entrapment. Because the isochores are near-vertical in the temperature range of greatest interest to petroleum geologists (50° to 200°C), Th does not greatly underestimate the true entrapment temperature. In order to determine the temperature of entrapment more accurately than this, one must be able to estimate the pressure of entrapment, which is fraught with difficulties. In order to use Th measurements to determine the crystallization temperature of the mineral phase, the fluid inclusions (1) must have been trapped initially as a single phase (ruling out the use of inclusions containing gas), (2) must be "primary" and (3) must not have stretched.

SALINITY (Tm)

The measurement of fluid inclusion salinity is an important way to discriminate meteoric, marine and hypersaline parent fluids. Salinity is important not only for determining the type of fluid which precipitated a mineral, but also as a parameter which can be used to identify families of fluid inclusions and correlate them with inclusions in other samples within the basin. Th and Tm can be used together to identify fluid inclusions which have stretched or leaked during burial.

Salinity is determined by first freezing an inclusion (using liquid nitrogen), and then raising its temperature in 0.5°C increments and recording the temperature at which it fully melts.

The final melting temperature, or temperature at which the last ice crystal in a frozen inclusion disappears, is a measure of the amount the freezing point of inclusion water is depressed below 0°C by the presence of dissolved salts. In general, more negative Tm values indicate higher salinities. However, the actual salinity cannot be estimated without making assumptions about brine composition because equal concentrations of different ions depress the freezing point by different amounts. By convention, salinity is calculated from Tm by assuming the fluid is a pure NaCl brine. Salinities are then expressed as wt% NaCl equivalents. The following table provides conversions from Tm to salinity from 0°C to the NaCl-H_2O eutectic at -20.8°C:

Tm (°C)	Salinity[6]	Tm	Salinity
0.0	0.0	-11.0	15.0
-1.0	1.7	-11.5	15.5
-1.5	2.6	-12.0	16.0
-2.0	3.4	-12.5	16.5
-2.5	4.2	-13.0	17.0
-3.0	4.9	-13.5	17.5
-3.5	5.7	-14.0	17.9
-4.0	6.4	-14.5	18.4
-4.5	7.2	-15.0	18.8
-5.0	7.8	-15.5	19.2
-5.5	8.5	-16.0	19.6
-6.0	9.2	-16.5	20.0
-6.5	9.9	-17.0	20.4
-7.0	10.5	-17.5	20.8
-7.5	11.1	-18.0	21.2
-8.0	11.7	-18.5	21.6
-8.5	12.3	-19.0	21.9
-9.0	12.9	-19.5	22.3
-9.5	13.5	-20.0	22.7
-10.0	14.0	-20.5	23.0
-10.5	14.5	-20.8	23.2

Most brines contain more than Na and Cl, and therefore the Tm values cannot be converted to *true* salinities by using the above table. The curves shown in Fig. 30 illustrate that Tm varies with both fluid inclusion salinity and composition. The composition of the major salts within a fluid inclusion can be obtained by the measurement of the temperature of first melting, known as the eutectic temperature, or Te. Te values for the important salt-water systems considered most relevant to aqueous fluid inclusions (from Shepherd, et al., 1985; Crawford, 1981; Haynes, 1985) are:

Salt System	Te (°C)
H_2O-NaCl	-20.8
H_2O-$MgCl_2$	-33.6
H_2O-NaCl-$MgCl_2$	-35.0
H_2O-$CaCl_2$	-49.5
H_2O-$CaCl_2$-$MgCl_2$	-52.2
H_2O-NaCl-$CaCl_2$	-55.0
H_2O-NaCl-$CaCl_2$-$MgCl_2$	-57.0

Te, which is difficult to observe, can only be measured in larger inclusions (which are rare in dolomite). In the literature, Te is rarely reported for carbonate minerals. The practice has been to convert Tm data to salinities for Tm values down to -20.8°C, and below this temperature report only Tm.

[6]In wt % NaCl equivalents.

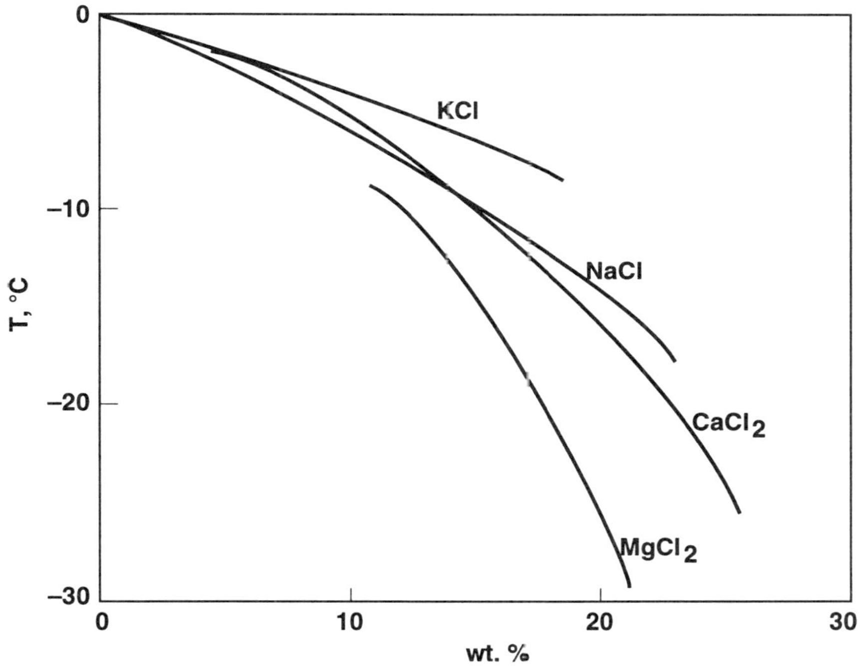

Crawford (1981)

Figure 30. The most easily observed phase transition during the melting of an aqueous fluid inclusion is the final melting temperature of the "last ice", or Tm. Salinity can be calculated from Tm if the composition of the inclusion fluid is known. If the fluid is known to contain predominantly Na and Cl, then, using this diagram, a Tm of -10°C would indicate a salinity of approximately 15 wt% NaCl equivalents. A $MgCl_2$ brine with a Tm of -10°C would have a salinity of approximately 11 wt% $MgCl_2$ equivalents.

In carbonate systems, the aqueous inclusion is often dominated by Na and Cl because most carbonates start out with their pores saturated with seawater or some derivative of seawater having these as the major ions. Contributions of Ca and Mg increase in evaporite-bearing sections where these cations are concentrated in the residual fluid by the evaporation of seawater and the precipitation of evaporite minerals, and are then expelled from the pore spaces of bedded evaporites during compaction. Thus, fluid inclusions from evaporite-bearing sections sometimes follow the $CaCl_2$ or $MgCl_2$ freezing point depression curves rather than the NaCl curve (Fig. 30). However, because the temperature of first melting is difficult to establish in the small inclusions which typify carbonates, the composition of the inclusion is rarely established, and it has become customary in carbonate studies to measure only the temperature of last ice and assume a NaCl-dominated fluid. This does not lead to significant errors in most cases, because most sedimentary basin brines are Na-Ca-Cl brines and the NaCl and $CaCl_2$ freezing point depression curves are similar, but in intervals adjacent to halite and potash evaporites, where residual brines are enriched in $MgCl_2$, salinity can be significantly overestimated by using the NaCl curve (Fig. 30).

Th-Tm PLOTS

It is instructive to crossplot fluid inclusion Th and Tm data. Such plots reveal a wealth of information concerning the history of an inclusion subsequent to trapping (Roedder, 1981).

Following entrapment in a subsiding basin, fluid inclusions are often deeply buried. When an inclusion is buried and heated, its internal pressure-temperature conditions change along the isochore appropriate to its density (as in Fig. 29). During burial, the pressure within an inclusion increases at a rate which is drastically higher than the rate of increase in the surrounding lithostatic and hydrostatic load (Fig. 31). Burruss (1987) calculates that an average inclusion will undergo an internal pressure increase of 400 bars above that of the surrounding hydrostatic pore pressure exerted on the host mineral when subsidence to an additional 1.5 km depth follows entrapment. Under these conditions, the fluid inclusion is overpressured, and the fluid cavity will stretch, or even hydrofracture. When simple stretching occurs, the inclusion's fluid does not leak, but the P-T conditions change due to an increase in cavity volume. This produces an increase in measured Th. When hydrofracturing occurs, the fluid in the inclusion leaks out and exchanges with the ambient pore fluid. This produces a new Th and Tm.

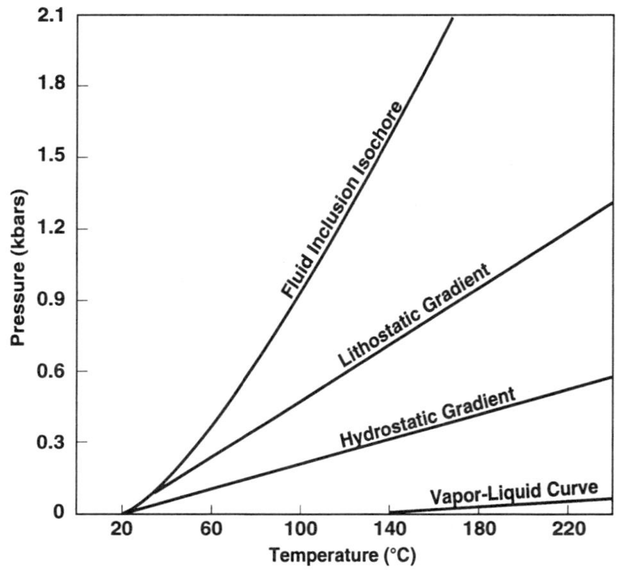

Goldstein (1986)

Figure 31. When rocks and interstitial fluids are buried, they are subjected to increasing temperatures and pressures along either a hydrostatic geothermal-geopressure gradient, a lithostatic geothermal-geopressure gradient, or somewhere between the two. In contrast, fluid inclusions, when buried and subjected to geothermal heating, act as a closed system and experience a much greater rate of pressure increase than the surrounding rocks and fluids. The resultant internal overpressuring will irreversibly expand (stretch) the fluid inclusion if the cavity wall has low strength (carbonates), and will hydrofracture the inclusion if the cavity wall is brittle.

The amount of stretching, which governs the new Th value for an inclusion, and the sequence in which individual inclusions within a population stretch, are not uniform. The smaller an inclusion, the larger the amount of overheating it will withstand before stretching begins. According to Burruss (1987), in smaller inclusions, 60°C of overheating will cause only a 10°C increase in Th. Fluid inclusions which contain a gas bubble (usually a hydrocarbon trapped as a discrete phase) are prone to leaking and stretching due to high internal pressures which are generated with increasing burial.

On a Th-Tm diagram, data collected on samples that have not been subjected to stretching or to leaking and re-equilibration plot in a circular "bullseye" (Fig. 32A). Samples that have been subjected to stretching show great variability in Th, but invariant Tm (Goldstein and Reynolds, 1988). Not only are Th values highly variable in a stretched population, the low-temperature end of the data set may also be shifted toward higher temperatures as well (Fig. 32B).

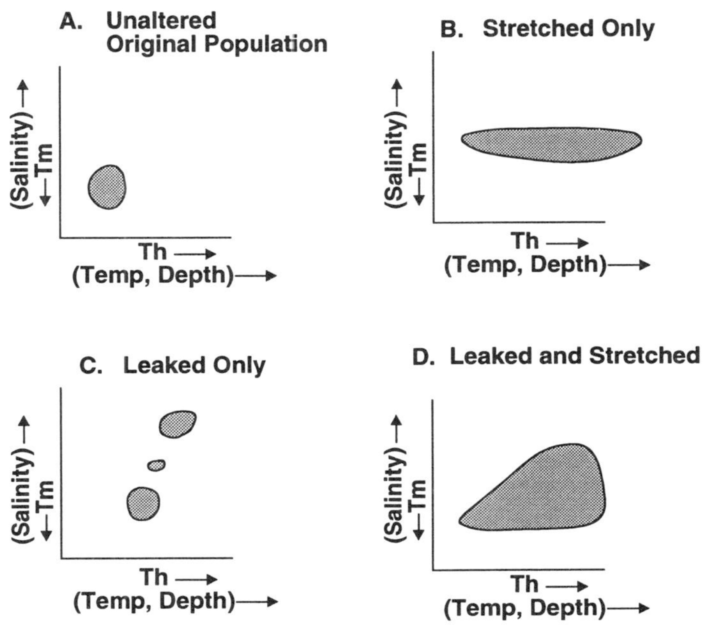

Modified after Goldstein and Reynolds (1988)

Figure 32. On a standard Th-Tm crossplot, data collected on samples in which fluid inclusions are unaltered appear as a tight cluster, as shown in (A). In samples containing fluid inclusions that have stretched but not leaked, the data appears as a wide spread of Th values with little variability in Tm (B). Fluid inclusions in (C), which have leaked through hydrofracturing and have not stretched, appear as discrete sub-populations. Since all inclusions do not leak at the same instant in time under identical conditions, the Th and Tm data may be highly variable, or may show a covariant increase in Th and Tm (because both temperature and salinity generally increase with depth of burial in sedimentary basins). When a fluid inclusion population re-equilibrates by both stretching and leaking as shown in (D), the data will appear as a wedge-shaped area, constrained by the leaking and stretching trends, which generally converge on the primary entrapment conditions.

Inclusions that leak and re-equilibrate generally trace a linear trend towards higher Th and Tm values (Fig. 32C), because formation waters commonly increase in temperature and salinity during progressive burial in a basin. As a result, inclusions that leak and re-equilibrate late in the burial history of a basin tend to be hotter and more saline than those which re-equilibrate early.

Finally, mixed populations composed of inclusions which have stretched and inclusions which have leaked appear as a wedge-shaped data set constrained by two lines which converge downward upon the original Th and Tm conditions (Fig. 32D).

Because only a part of the fluid inclusion population re-equilibrates during burial, and because re-equilibration is a stepped process which is dependant on inclusion size, shape, and proximity to cleavage traces, fluid inclusion crossplots may record several episodes in a basin's history. In subsiding basins, strata are subjected to increasing temperatures with burial. In this setting, formation-fluid and fluid inclusion salinities may increase along with temperature. Other basins experience periods of uplift and/or erosion, which can be recorded as a generation of late, meteoric cement containing inclusions with characteristically low Th and Tm values. Tectonic and thermal events can also be recorded in fluid inclusion populations. Therefore, fluid inclusions provide valuable information on basin evolution. The application of these techniques to the analysis of basin history in England, Dubai, the U.S. Gulf coast, and offshore Norway is nicely illustrated in McLimans (1987).

STRONTIUM ISOTOPES

THEORY

There are four isotopes of Sr. Their approximate proportions are: ^{84}Sr (0.56), ^{86}Sr (9.87), ^{87}Sr (7.04) and ^{88}Sr (82.53). ^{87}Sr and ^{86}Sr are similar in abundance, and their ratios are of practical significance in the study of formation waters and carbonates.

These two isotopes of Sr originate from separate sources (Fig. 33; Faure, 1977; Faure and Powell, 1972). ^{86}Sr is non-radiogenic, whereas ^{87}Sr is radiogenic and forms from the radioactive decay of ^{87}Rb, which has a half-life of approximately 50 billion years.

The primordial $^{87}Sr/^{86}Sr$ ratio incorporated into the Earth at its formation was 0.699 (see Veizer, 1989). Fractional crystallization of the Earth's crust has resulted in high Rb content of the continents because singly charged Rb substitutes easily for K in potassium feldspars and clay minerals, both of which are abundant in old sialic crustal rocks (e.g., granites, gneisses and arkoses). The Rb content in the residual mantle material has diminished as a result of this continental enrichment. The mantle's low Rb content is reflected today in the low $^{87}Sr/^{86}Sr$ ratio (approximately 0.705) of simatic rocks, whereas the continents' high Rb content has increased the $^{87}Sr/^{86}Sr$ ratio of sialic rocks to more than 0.716, on average (Fig. 34; and Veizer, 1989). Sr with a low 87/86 ratio is found in abundance as a minor or trace element substituting for Ca in plagioclase feldspars, amphiboles and pyroxenes, which are more common in undifferentiated basic rocks. It also substitutes for Ca in carbonate minerals.

The dichotomy between the mantle and the continents is important in understanding the historical changes in the Sr isotopic composition of seawater and marine carbonates (Brass, 1976). This is because the Sr isotopic composition of the oceans is a balance between continental runoff and groundwater runout on the shelves (both of which are sources of continental Sr), and the interaction of seawater with mid-ocean ridge basalts and with water from submarine vents (which act as sources of mantle Sr). The Sr isotopic composition of the oceans is controlled by the balance between (1) erosion of igneous and metamorphic, sialic rocks of the old shields, which provides radiogenic Sr to seawater, and (2) erosion of volcanic rocks on the continents and alteration of basalts at mid-ocean ridges, which provide non-radiogenic Sr to seawater. Large rivers with extensive drainage basins that contain both acidic and basic rocks deliver Sr of an intermediate isotope ratio to the oceans.

As a result, the Sr isotopic composition of seawater has varied through time as a function of the relative fluxes of continental and mantle Sr. The continental flux is highest during times of orogenesis, whereas the mantle-flux is highest during times of rapid sea-floor spreading. The high flux of Sr into the oceans ($5.4X10^{12}$ g/y), combined with the long oceanic residence time of Sr (>4 my) and the rapid mixing rate of the oceans (10^3 yrs), have caused the Sr isotope ratio of seawater to be globally homogeneous at any given time and temporally sensitive to climate and tectonics (Veizer, 1989).

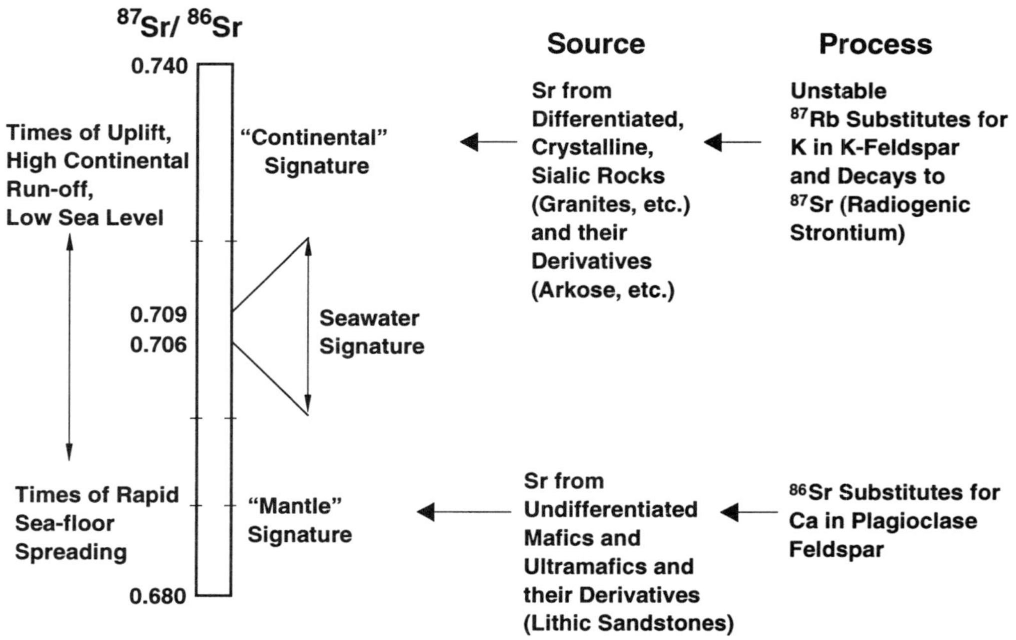

Figure 33. Summary of the sources of ^{87}Sr and ^{86}Sr and the paths which influence the overall Sr isotopic composition of paleo-oceans. Note that mountain building and erosion release Sr having a high ^{87}Sr/^{86}Sr ratio into the oceans (generally corresponding to sealevel lowstands). Mid-ocean ridge activity (generally corresponding to times of sealevel highstand) releases Sr having a low ^{87}Sr/^{86}Sr ratio into the oceans.

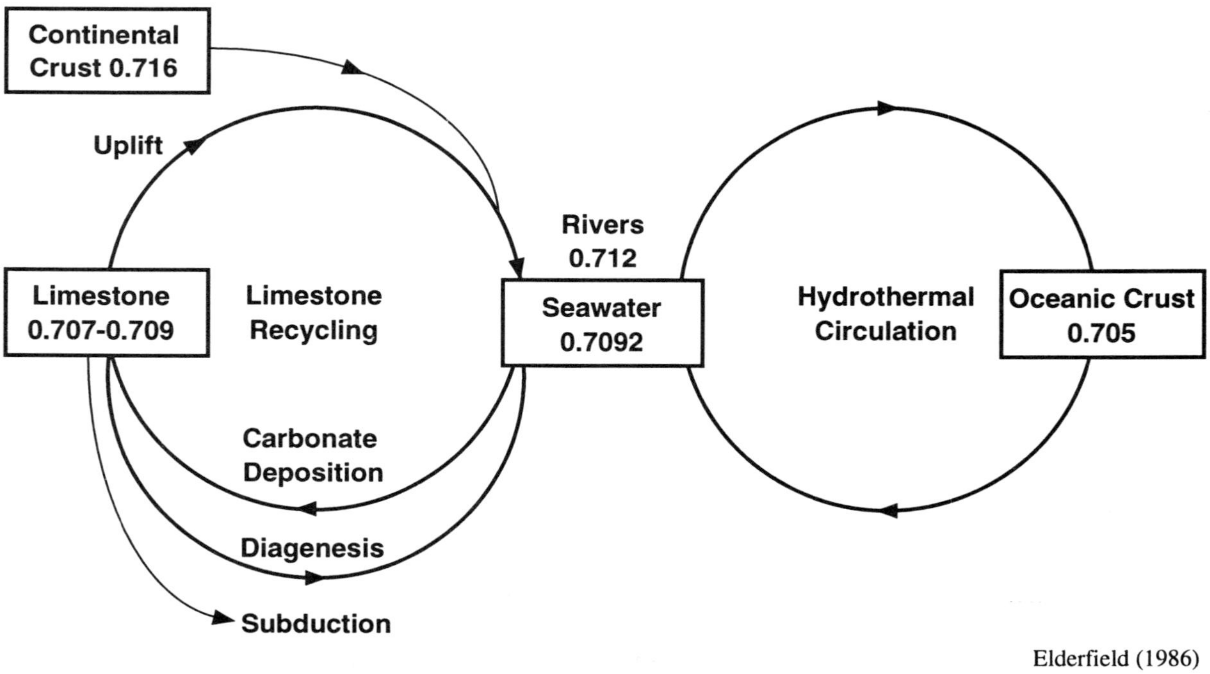

Figure 34. This diagram shows the relationships between the important reservoirs for radiogenic and non-radiogenic Sr and the geochemical cycles that exert control over the Sr isotopic composition of seawater.

APPLICATIONS

In carbonate studies, Sr isotopes are used most often as an age-dating tool. The Sr isotopic composition of ancient seawater and its variation through geologic time have been determined from marine carbonate minerals. Secular changes in the $^{87}Sr/^{86}Sr$ ratio of seawater are the result of the interplay between sea floor spreading (high sea floor spreading rates decrease the $^{87}Sr/^{86}Sr$ ratio) and diastrophism and erosion (increase in tectonism and the resultant erosion increase the $^{87}Sr/^{86}Sr$ ratios). The $^{87}Sr/^{86}Sr$ ratio of a carbonate mineral can be measured with great precision, to five significant figures (0.0000X). To trace the change in Sr isotopic composition of seawater through the Phanerozoic, Burke, et al. (1982) constructed a reference curve based on the $^{87}Sr/^{86}Sr$ ratios of hundreds of marine carbonates (Fig. 35). The ages of the samples used to construct this curve were constrained by index fossils. This curve serves as a reference for age-dating marine carbonates lacking in diagnostic index fossils. Sr isotope ratios are particularly useful for dating parts of the Cenozoic. The lower and middle Miocene section can be dated to within ± 0.3 my using this curve (DePaolo, 1986; Finger and DePaolo, 1989). In more recent time intervals, the availability of fossils in good condition permitted construction of very detailed segments on the Sr isotope age curve. In some cases the technique yields a higher stratigraphic resolution than biostratigraphy. In more ancient time intervals, poor fossil preservation and imprecise age dates have not permitted the control necessary for an accurate Sr isotope age curve.

The Sr isotope age curve displays an oscillating trend having a large amplitude, and a high frequency caused by short-term climatic, tectonic-erosion, and seafloor-spreading cycles. The calibration points along this curve are determined by biostratigraphy. One should be extremely careful when interpolating between these calibration points, particularly when dealing with time intervals characterized by high frequency oscillations with periods less than the biostratigraphic resolution, because the sedimentary record is composed largely of units deposited at variable rates which are broken by prolonged, poorly-defined periods of non-deposition and erosion. The Sr isotope age curve is most useful for dating time intervals in which the curve is steep and unidirectional. It has been used with particular success in the lower and middle Miocene and the middle and late Permian.

Sr isotope ratios which depart from the marine baseline can be used to discriminate lacustrine carbonates because their isotope ratios reflect that of the continental watershed. This is important in rift basins where timing of marine incursions may be difficult to establish by other means.

Sr isotopes can also be used to constrain the timing and environment of formation of Cenozoic dolomites. Saller (1984) and Swart, et al. (1987) proposed seawater convection models for Cenozoic dolomite from Enewetak Atoll and the Bahama Platform, respectively, using Sr isotopes combined with other evidence.

Sr isotopes serve as tracers of subsurface fluid movement (Burtner, 1987), specifically fluid which has interacted with siliciclastics or crystalline basement. Oil field brines usually contain excess ^{87}Sr derived from interaction with aluminosilicates (potassium feldspar and illite), although brines which have flowed through young volcaniclastic sediments may contain excess ^{86}Sr (Schultz, et al., 1989). Sr released by diagenetic processes is partitioned into carbonate cement and dolomite forming in the

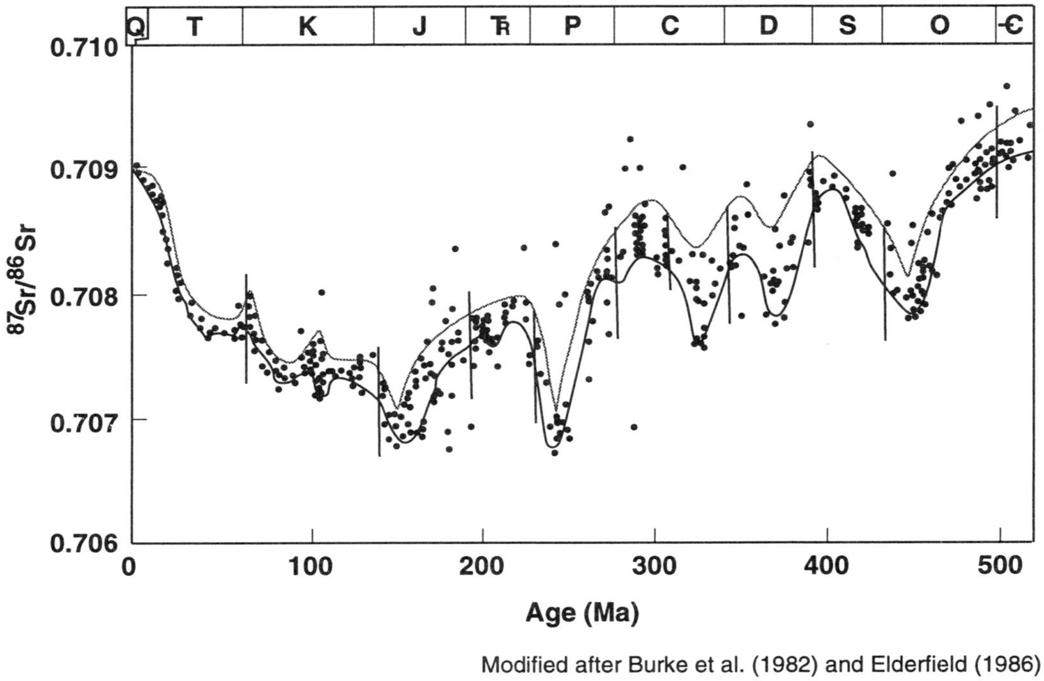

Figure 35. Strontium isotope age curve based on the $^{87}Sr/^{86}Sr$ ratios of marine carbonate rocks. Curve provides a best estimate of the $^{87}Sr/^{86}Sr$ ratio of seawater as a function of geologic age. Higher Sr isotope values (top part of the data envelope) may reflect addition of radiogenic (high $^{87}Sr/^{86}Sr$ ratio) Sr contaminants through diagenetic alteration; therefore, the best estimate for the true seawater value is closer to the lower side of the data envelope (solid line). The accuracy of this trend line is limited by the resolution of the biostratigraphic age determination for each sample. The data therefore appear as a band of increasing width with increasing age. Oscillations in the Sr isotope signal of seawater are better defined for the Cenozoic and Mesozoic where stratigraphic resolution is best. Oscillations in the post-Paleozoic may be linked to the break-up of Pangaea, with radiogenic excursions corresponding to continental erosion, and non-radiogenic excursions to increases in rates of seafloor spreading.

subsurface because Sr, which is a doubly-charged cation, substitutes easily into the carbonate lattice. The weathering of illite and potassium feldspar, albitization, plagioclase-feldspar dissolution, and conversion of smectite to illite causes the $^{87}Sr/^{86}Sr$ ratio of pore water to vary widely. When a pore fluid's Sr isotopic composition differs significantly from that of the dolomite, dolomitization preceded emplacement of the fluid; on the other hand, where the fluid matches the dolomite, dolomitization is likely to have occurred relatively recently, from the current fluid.

TRACE ELEMENTS

THEORY

There is extensive literature that deals with trace elements in carbonate minerals. However, trace elements do not always yield unambiguous information with reasonable amounts of effort. The time and effort that must be expended to collect quantitative trace element data is significant, and includes the operation of an electron microprobe, an atomic absorption spectrophotometer and/or an inductively-coupled plasma spectrometer (ICP). Two minor or trace elements that are of very practical significance and, fortunately, can be readily detected (but not quantified) using stains and cathodoluminescence microscopy (CL), are Fe and Mn; they are discussed in this section. Sr and Na have also been used extensively, but with less success. We recommend a detailed, quantitative analysis of other minor and trace elements only when objectives are clearly defined and an investment of time and energy is warranted.

Veizer (1983) discusses trace elements that are important in carbonate minerals. The concentration of a minor or trace element in a dolomite, as in other minerals, is governed by the element's Distribution Coefficient and its abundance in the parent fluid. This is the simple case in which equilibrium between the mineral and fluid is achieved. When the precipitation rate is very fast (disequilibrium), then the trace element concentration in the mineral approaches that of the fluid; this is known as a "kinetic effect".

The most common way minor and trace elements are incorporated into carbonate minerals is in substitution for Ca^{+2} in the carbonate lattice. The concentration of an element in a carbonate mineral is governed by the element's Distribution Coefficient, which is defined as follows:

$$[m_{Me}/m_{Ca}]_{(solid)} = \mathbf{D} * [m_{Me}/m_{Ca}]_{(water)}$$

where "m" indicates molar concentrations, "Me" stands for the trace element (metal), and D is its Distribution Coefficient. Simply stated, this means that the concentration of an element in a mineral is directly proportional to its concentration in the water.

The Distribution Coefficient of a minor or trace element within a carbonate mineral will approach 1.0 as its ionic radius approaches that of the major element for which it substitutes, in this case, calcium. Elements which are larger than Ca will not fit as easily into the lattice as will elements which are smaller than Ca. Ionic radii (for 6-fold coordination) of elements that commonly substitute in the dolomite lattice *for Ca* are shown below:

Element	Ionic Radius	Dist. Coeff., for substitution for Ca in Dolomite
Sr	1.13Å	0.026 to 0.06
Ca	1.00Å	1.00
Mn	0.83Å	5-30
Fe	0.78Å	1-20

This table shows that Sr, which is larger than Ca, remains behind *in the water* during dolomitization, whereas elements (e.g., Fe and Mn) which are smaller or closer to the ionic radius of Ca fit more easily into Ca sites and thus are concentrated *in the mineral*. A similar table could be constructed for elements substituting for Mg in dolomite. Mg has an ionic radius of 0.72Å, which permits substitution by both Mn and Fe, but not Sr. Ankerite,

$$[Ca(Mn,Fe)]CO_3\,[Mg(Fe,Mn)]CO_3$$

is an example of substitution of Mn and Fe in both the Ca and Mg positions of a dolomite structure.

Because Fe and Mn have D >1 and Sr has D <1, the dolomitization of limestone will decrease the concentration of Fe and Mn *in solution* and increase the concentration of Sr. This will lead to concentration gradients characterized by decreasing Fe and Mn and increasing Sr along fluid flow paths. These regional gradients in fluid chemistry should be reflected in the minor and trace element content of dolomite formed at various locations along a flow path. Buelter and Guillemette (1988) mapped regional concentration gradients in Fe, Mn, and Sr *in dolomite* in the Bonneterre Formation (Cambrian), southeast Missouri and used trends of decreasing Fe and Mn coupled with increasing Sr to determine the direction of flow of the fluid which dolomitized the Bonneterre.

The second important factor in determining the concentrations of minor and trace elements in dolomite is their concentrations in the parent fluid. For example, Sr content is high in seawater and formation fluids but low in fresh water, so low Sr concentrations in dolomite are sometimes taken to indicate dolomitization in mixed marine-meteoric waters. The same holds true for Na. The use of Sr and Na is fraught with difficulties, however since early marine and evaporitic dolomites may lose some of their Sr and Na if flushed with fresh water during exposure. In addition, much of the Na may reside in fluid inclusions rather than in the carbonate lattice, and is therefore not governed by the equation for the Distribution Coefficient. Furthermore, uptake of Sr and Na in carbonates is particularly sensitive to growth rates (kinetic effects); the slower the growth rate, the more effective the exclusion of Sr and Na.

Fe and Mn can be used more successfully than Sr and Na to differentiate dolomitization environments, by virtue of the fact that Fe and Mn are present in low concentration in seawater, but relatively abundant in formation fluids (Veizer, 1983). The ratios Fe/Ca and Mn/Ca are, on the average, higher by a factor of 1,000 in formation fluids than in seawater:

Element [ratio]	ppm or [Ratio]	
	seawater	oil field fluid
Ca	411.	1,000-20,000
Fe	0.002	0.01-500
Mn	0.0002	0.1-100
[Fe/Ca]	[10^{-6}]	[10^{-3}]
[Mn/Ca]	[10^{-7}]	[10^{-4} to 10^{-3}]

This means that dolomite precipitated in the subsurface may be enriched in Fe and Mn compared to marine dolomite by greater than three orders of magnitude. Most marine dolomite contains less than 50 ppm Mn and Fe, whereas subsurface dolomite may contain as much as 150,000 ppm Fe and 10,000 ppm Mn (grading into ankerite). Thus, high abundance of Fe and Mn provides an effective way to identify burial dolomite.

Only the reduced form of Fe (Fe^{+2}) is incorporated into the carbonate mineral; therefore, a second important reason surface dolomite lacks Fe is that it is precipitated under oxidizing conditions in which Fe is in the oxidized state (Fe^{+3}). Low concentrations of Fe can, however, occur in subsurface dolomite where the pore fluid is "sour" due to high concentrations of H_2S. In this case, most of the Fe is removed from solution as FeS_2 and is therefore unavailable for incorporation into dolomite.

Without resorting to trace element analyses, the petroleum geologist can determine whether a dolomite is enriched in Fe or Mn with simple petrographic techniques, which are discussed next.

CATHODOLUMINESCENCE AND STAINING TECHNIQUES

Cathodoluminescence (CL) is the emission of light resulting from the bombardment of, in the case of carbonates, Mn using a cathode ray (Frank et al., 1982). The process is similar to what happens in the electron microprobe, except, instead of inducing X-ray emission with a high energy electron beam, the luminoscope induces visible light emission using a low energy electron beam. In the luminoscope, the unfocused beam bumps an electron out of an orbital shell within the Mn atom, whereupon the electron, in returning to its stable position, emits light.

Whereas Mn is the source of luminescence, Fe retards or "quenches" luminescence (Fig. 36). The strength of the luminescence is independent of the absolute concentration of Mn, but instead depends on the Mn/Fe ratio in the dolomite, with the effects of Fe and Mn tending to counter each other. Small changes in this ratio have been observed to produce large variations in luminescence within a single crystal. This can result in exquisite CL zonation within dolomite cements (e.g., see Cander, et al., 1988). Variations in the Mn/Fe ratio are caused by changes in the supply of Fe and Mn to the pore water, in the oxidation state of the pore water, and in the H_2S content of the pore water (which governs the availability of reduced Fe for precipitation in carbonate). The luminescence response to Eh and pH is shown in Figure 37. Since small variations in Eh can produce large changes in luminescence response, Eh is the more important of the two factors. Recent studies have identified many factors other than Eh, pH, and trace element concentration in the fluid which play roles in trace element partitioning and CL zonation (e.g., crystal growth rate, sector zoning, and bulk fluid disequilibrium). Machel and Burton (1991) provide an in-depth discussion of the complexities involved.

Another simpler way of analyzing dolomite for presence or absence of Fe is to apply a K-ferricyanide stain to the uncovered surface of a thin section or a polished slab (Dickson, 1966). Fe-rich dolomite will stain blue, Fe-poor dolomite will not stain. Staining with K-ferricyanide will often reproduce the exquisite zonation revealed by the CL technique, with the heavily stained layers correlating with non-luminescing mineral. K-ferricyanide staining is often combined with Alizarin Red S staining, in order to identify Fe-rich calcite (purple) and Fe-poor calcite (red). Staining is much less time

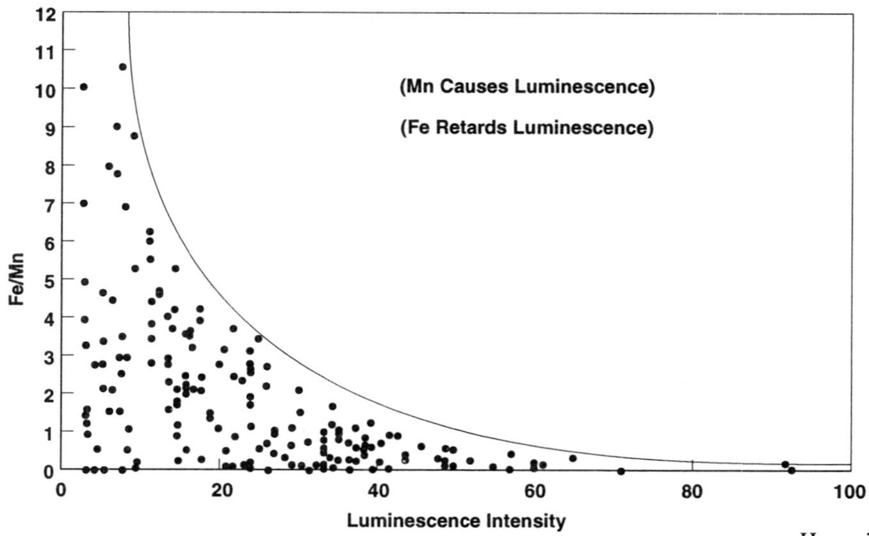

Hemming, et al. (1989)

Figure 36. The intensity of luminescence within carbonate minerals that are subjected to bombardment with an electron beam ("cathodoluminescence") is inversely proportional to the Fe/Mn ratio. Mn excites luminescence, whereas Fe retards or "quenches" luminescence. Luminescence intensity on the X-axis is measured on a spectrophotometer and scaled to 100.

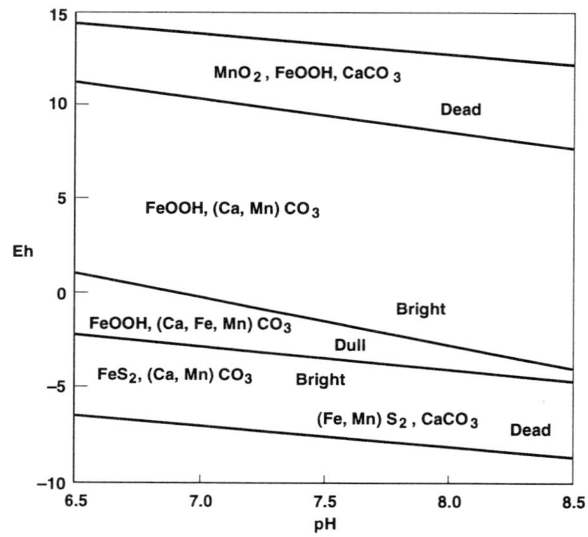

Frank, et al. (1982)

Figure 37. The oxidation state (Eh) and concentration of Mn and Fe in a pore fluid are the main controls on the Mn and Fe content of a carbonate mineral and, thus, its luminescence. Pore fluids with a high Eh, in which both Mn and Fe are in the oxidized form (MnO_2 and FeOOH), precipitate non-luminescing ("dead") carbonate (top of diagram). Where the Eh of the fluid is somewhat lower, only Mn may be divalent (Mn^{+2}) and capable of being incorporated into carbonate. The resulting manganoan carbonate luminesces brightly. In the case where the Eh of the fluid is still lower, some, but not all, of the Fe is reduced and partitioned into the carbonate, causing the luminescence intensity of the carbonate to be "dull". In the case where Eh is very low and all Fe and Mn are divalent, the carbonate will have a higher Fe/Mn ratio and may not luminesce. However, in sour waters, Fe can be precipitated as FeS_2; carbonates precipitated from these waters will be manganoan and non-ferroan, and will luminesce brightly.

consuming than CL microscopy, and can be performed on slabbed cores more easily than can CL microscopy.

In summary, the presence of Fe and Mn is significant because it indicates that a dolomite formed in the subsurface (sabkha and supratidal dolomites are devoid of Fe and Mn). Fe can be detected in dolomite using K-ferricyanide stain, and Mn can be detected using cathodoluminescence microscopy.

PART II:
CASE STUDIES

In Part II of this manual, we present case studies of reservoirs that formed in each of the major dolomitization environments. Supratidal-sabkha, reflux, marine-meteoric mixing zone, and burial dolomite reservoirs are each discussed in separate sections. The examples were selected to best illustrate the ways in which geochemistry was used successfully in conjunction with stratigraphic, structural, petrographic, and facies information to identify the origin of the dolomite reservoir. We believe that these examples will serve as useful templates for the explorationist who wishes to use geochemical data to evaluate dolomite petroleum reservoirs and to determine which dolomite model best fits his or her reservoir or play. A summary of the geochemical criteria for recognition of each of these dolomite types is presented in Table 2.

Table 2. Variations in Geochemical Characteristics Due to Dolomitization Processes

GEOCHEMICAL CHARACTERISTICS

DOLOMITIZATION PROCESSES	Stable Isotopes		Fluid Inclusions			$^{87}Sr/^{86}Sr$	Trace Elements		
	$\delta^{18}O$	$\delta^{13}C$	Fluorescence	Th	Tm		Sr and Na	Fe and Mn	CL
Subtidal-Supratidal Cycles	≥Marine carbonate baseline	=Marine carbonate baseline	None	None (single phase)	Low to High	=Marine baseline for host sediment	High	Low	Dead
Reflux	≥Marine carbonate baseline	=Marine carbonate baseline	None	None (single phase)	Low	=Marine baseline for overlying evaporites	High	Low	Dead
Marine and Mixing Zone	≤Marine carbonate baseline	≤Marine carbonate baseline	None	None (single phase)	High	=Marine baseline for host sediment	High to Low (variable)	High to Low (variable)	Variable and zoned
Burial	<Marine carbonate baseline	≤Marine carbonate baseline	Possibly	>50°C	Low	Unequal to marine values	Variable	High	Dull

DOLOMITES ASSOCIATED WITH SUBTIDAL TO SUPRATIDAL CYCLES

Generation of hypersaline brines by evaporation of seawater near the surface of sabkhas and supratidal flats (Fig. 38) results in precipitation of anhydrite, which increases the Mg/Ca ratio of the residual brine (Fig. 39). The lateral and vertical movement of this brine results in dolomitization of supratidal and underlying intertidal and subtidal sediments.

Shoaling upward cycles topped by tidal flats or supratidal evaporites produce reservoirs that follow the shorelines of basins, lagoons, and epeiric seas. Repeated transgressions and regressions result in multiple wedges of subtidal and intertidal sediments beneath and within impermeable, often anhydritic supratidal facies (Fig. 40; Shinn, 1983). This produces stratigraphic traps with seals composed of supratidal anhydrite mixed with impermeable microcrystalline dolomite. Both overlie more permeable, more coarsely crystalline ("sucrosic") subtidal dolomite which forms the principal reservoir facies. Solution of evaporites can produce collapse breccias, which also make excellent reservoirs. Supratidal reservoirs tend to be relatively thin but can be laterally extensive.

RECOGNITION

Supratidal dolomite is usually stratiform and preserves sedimentary structures. In arid depositional settings it is associated with sedimentary features characteristic of the sabkha environment, e.g., algal mats, ripup clasts, nodular anhydrites, and evaporite solution-collapse breccias. Dolomitization can often be seen to precede compaction and may be paragenetically related to early diagenetic textures and cements. Sabkha dolomite is erratically distributed throughout a section and exhibits local rapid vertical and lateral shifts into limestone, evaporites, and siliciclastics. Because supratidal flats and sabkhas are subaerially exposed, they tend to be invaded by meteoric waters in their landward portions. This commonly results in dissolution of bioclasts, pellets, ooids, and evaporite crystals to produce moldic porosity in supratidal and underlying subtidal facies. Porosity in sucrosic dolomite can be further enhanced by removal of remaining undolomitized mud matrix. Crystal size is commonly <15 µm (microcrystalline) in supratidal facies, but often coarsens to 40-100 µm (very finely crystalline) in associated sucrosic subtidal facies. Evidence of preferential dolomitization of muddy matrix is often apparent.

Because sabkha dolomite forms syngenetically from evaporated seawater, **$\delta^{18}O$ compositions should be heavy, Sr and Na contents should be high, Mn and Fe contents should be low, and Sr isotopic compositions should be similar to values for seawater at the time of sediment deposition.** However, geochemical signatures can be reset by later recrystallization in invading fresh waters or during burial. Because supratidal and sabkha dolomites form at the surface, two-phase fluid inclusions are absent (Moore, 1989; Warren, 1989). **Single-phase fluid inclusions may be present and should display low Tm values indicative of high salinity,** unless resetting in fresh water has occurred.

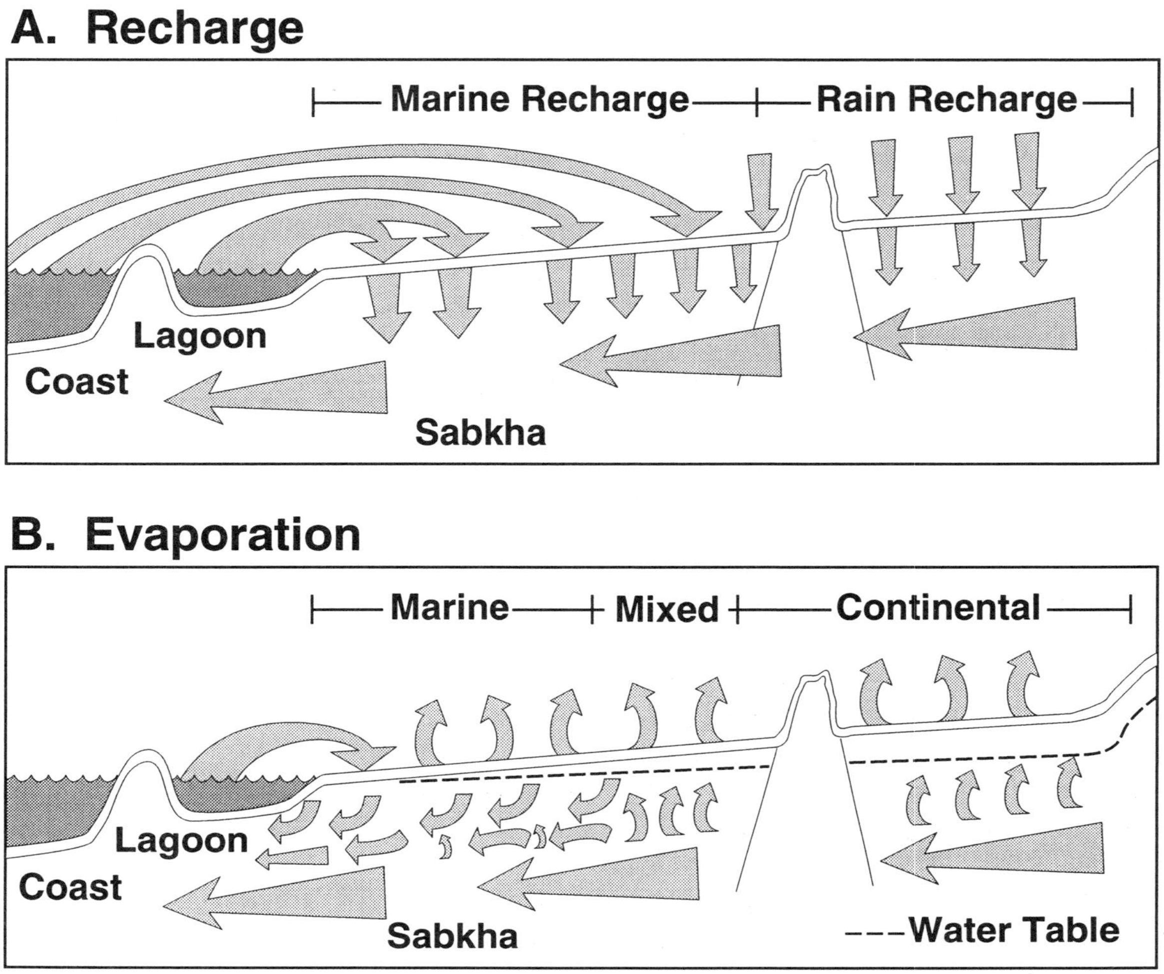

McKenzie and others (1980)

Figure 38. (**A**) Sabkha waters can be derived from periodic marine storm recharge, rain falling on the sabkha surface, or influx of continental waters from coastal aquifers. (**B**) During dry seasons, hypersaline brines are generated by evaporation of the complex fluid which results from intermixing the above waters. These brines are dense and flow downward and seaward. In some cases,they may move upward as well, due to capillary action.

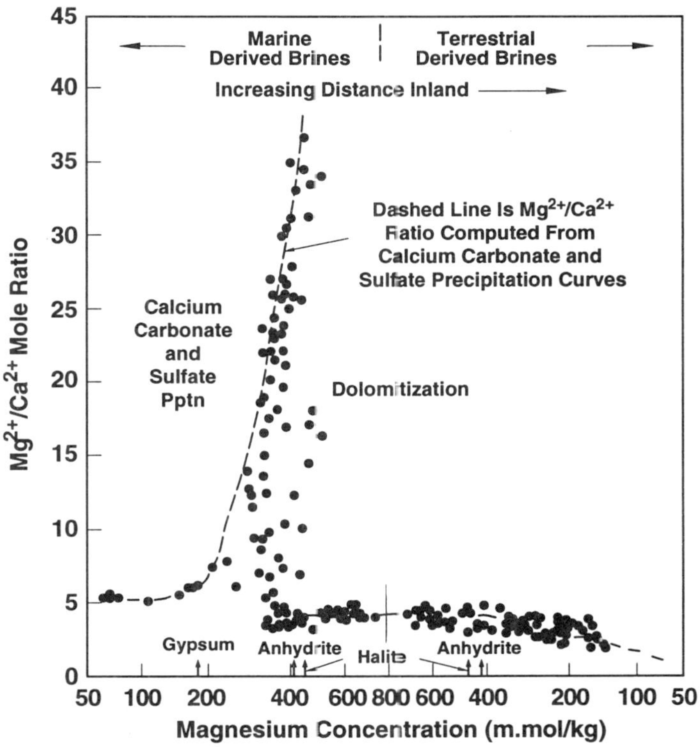

Butler (1969)

Figure 39. Evaporating seawater, which moves onto the sabkha from the left, progressively increases in salinity in a landward direction and initially precipitates aragonite and gypsum in sabkha sediments. Precipitation of these calcium-bearing minerals increases the Mg/Ca ratio of pore fluids to the point where dolomite begins to form, both as a replacement mineral and as a direct interstitial precipitate. Dolomitization of lime sediments brings the Mg/Ca ratio down again. Terrestrial waters enter the sabkha from the right and evaporate as well, but Mg/Ca ratio does not increase significantly. Variation in Mg/Ca ratio across the sabkha, the zone where dolomitization occurs, and the points at which gypsum, anhydrite, and halite first appear in the sediments (arrows) are indicated.

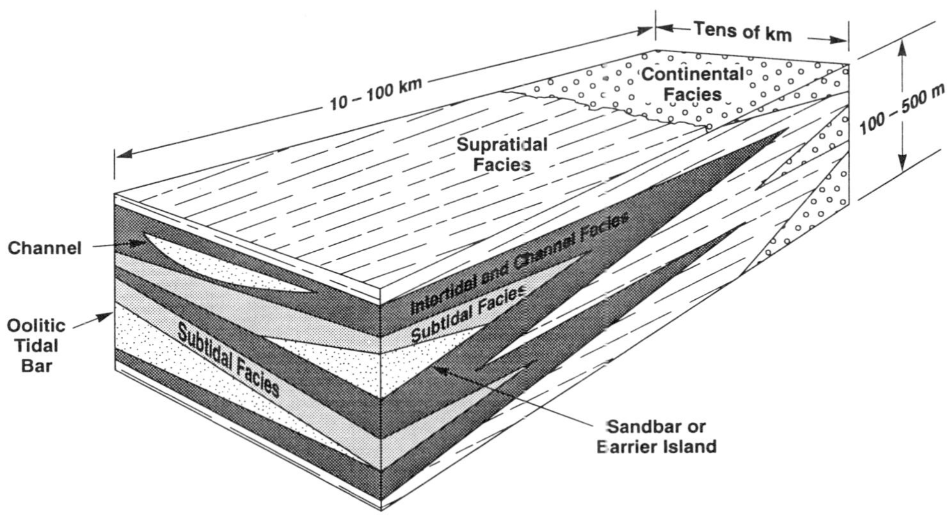

Shinn (1983)

Figure 40. Repeated transgressions and regressions produce wedges of subtidal and intertidal sediments which pinch out beneath and within impermeable, often anhydritic, supratidal sediments. Subtidal facies, which form the reservoir, are commonly composed of permeable, finely crystalline dolomite whereas supratidal facies, which form the trap, are composed of impermeable microcrystalline dolomite. Petroleum migration and accumulation are enhanced by down-to-the-basin tilting during burial.

RED RIVER FORMATION

Cabin Creek Field on the Cedar Creek Anticline in the southwestern portion of the Williston Basin (Fig. 41) produces hydrocarbons from the Upper Ordovician Red River Formation (Fig. 42). The field is in one local structural closure of about 250 ft along the 100 mi long Cedar Creek Anticline. The vertical seal is formed by tight dolomicrites and black shale stringers within the Red River cycles and by the overlying Stony Mountain Shale. Red River production is commingled with production from Silurian peritidal carbonates of the Interlake Formation. Combined primary recovery is estimated at 75 MMBO, with secondary recovery from waterflooding estimated at an additional 14 MMBO.

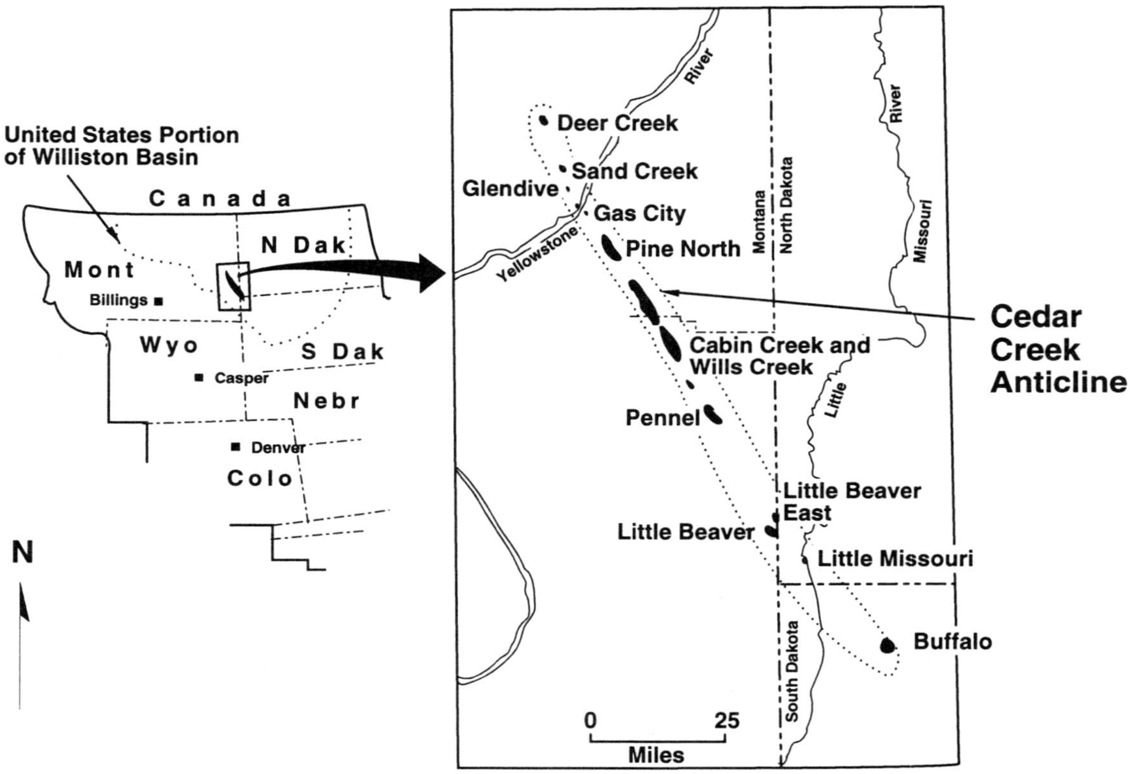

Roehl (1985)

Figure 41. Index map showing location of Cabin Creek Field on Cedar Creek Anticline.

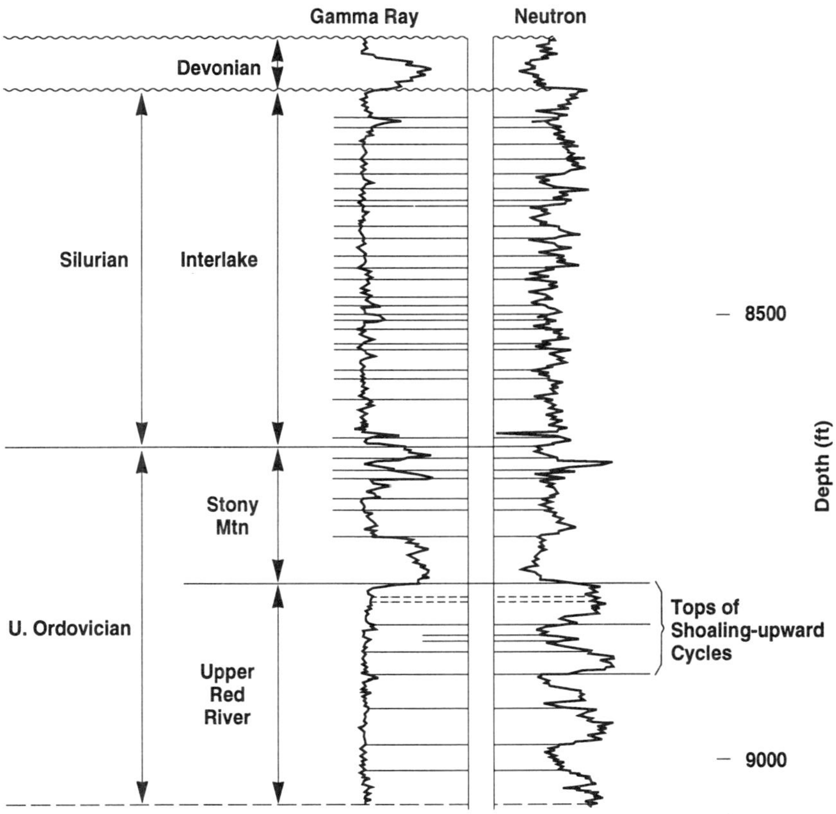

Modified after Roehl (1985)

Figure 42. Age, stratigraphic relationships, and gamma ray - neutron log character, Upper Red River, Stony Mountain, and Interlake formations, Cabin Creek Field.

The Red River Formation at Cabin Creek Field provides a clearcut "endmember" example of a supratidal dolomite reservoir with minimal downward reflux of brine. It was not necessary to use geochemistry to identify the dolomitization environment, in this case, since macro- and micro-scale petrographic criteria for identifying simple sabkha dolomite reservoirs are so straightforward.

Deposition of the reservoir interval proceeded on a carbonate ramp, with the updip portion characterized by mud-supported sediments with thin intercalated grainy beds. The upper 150 ft of the Red River is composed of three shoaling upward sequences topped by sabkhas. The upper surface of each sabkha is erosional (Roehl, 1985 and Fig. 43). Intertidal sediments contain ripple and cross lamination, imbricated shells, intraclasts, and oncolites, and grade upward into supratidal sediments containing desiccation cracks, wavy laminations (probable stromatolites), flat pebble breccias, and truncation surfaces. Both intertidal and supratidal sediments have been altered to micro- to extremely fine crystalline (10 to 60 µm) dolomite. Both contain abundant molds of gypsum and anhydrite. Supratidal sediments contain nodular anhydrites and evaporite solution-collapse breccias. Underlying subtidal units are skeletal and pelletal wackestones and packstones and are undolomitized in their lower portions. The lime mud matrix has been dolomitized in the uppermost portions of subtidal units, while relict grains are preserved as calcite or are leached.

Upper Red River Formation

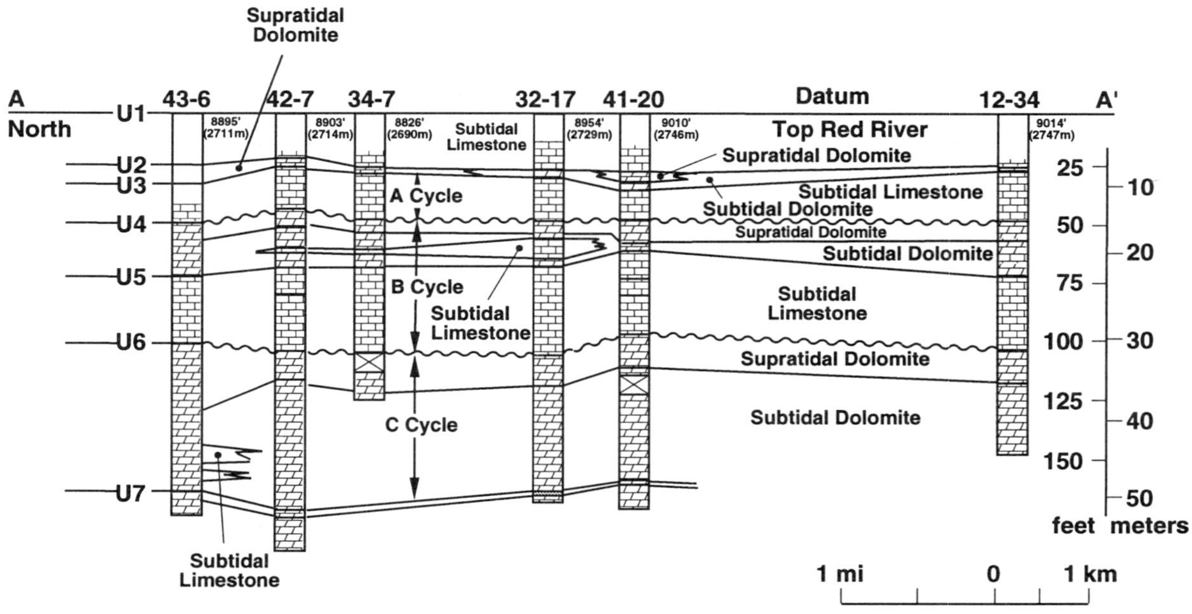

Modified after Ruzyla and Friedman (1985)

Figure 43. Northwest-southeast stratigraphic cross section along the eastern flank of Cabin Creek Field showing facies relationships within three shoaling-upward cycles (A,B,C) in the Red River Formation which are topped by sabkhas with erosional upper surfaces. U1 to U7 refer to log markers.

Dolomitization is interpreted as syngenetic because of fine crystal size, preservation of depositional constituents and textures which show no signs of compaction (implying dolomitization occurred early and arrested compaction) and preservation of molds and vugs with shapes that mimic gypsum and anhydrite crystals observed on modern sabkhas. The dissolution of evaporites in supratidal and intertidal facies and undolomitized calcite grains in subtidal facies is interpreted to have occurred during vadose exposure at the erosional surfaces. Porosity is predominantly diagenetic in origin and is restricted to supratidal, intertidal, and the uppermost parts of subtidal facies (Fig. 44). The two dominant porosity types are intercrystalline and moldic to vuggy. Such finely crystalline dolomite typically has low permeability. This can be seen in Fig. 44 where k rarely exceeds 10 md. Anhydrite has been remobilized and reprecipitated locally in moldic, intercrystalline, and interparticle pores. This is a common problem in supratidal reservoirs (Ruzyla and Friedman, 1985).

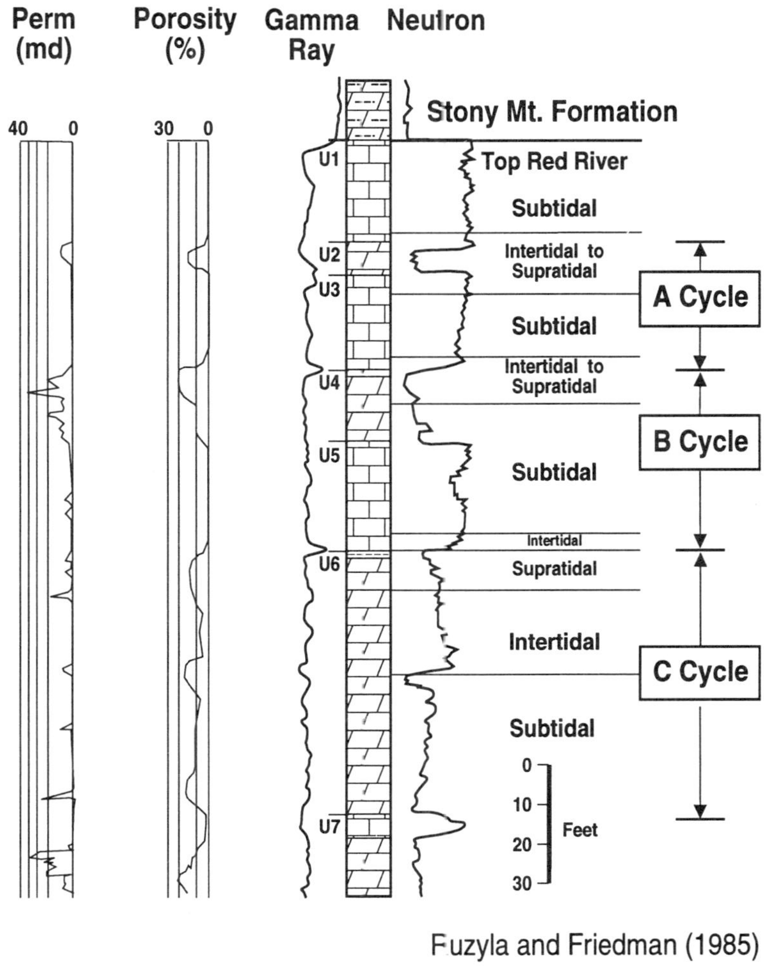

Figure 44. Relationships between neutron porosity log response, core porosity, core permeability, and depositional facies. Porosity was produced by supratidal dolomitization augmented by limited seepage reflux dolomitization in intertidal and subtidal facies. Dolomite reservoir is confined to the supratidal, intertidal, and upper part of subtidal facies.

SAN ANDRES FORMATION

Subtidal to supratidal cycles which contain dolomite reservoirs that pinch out updip into tight supratidal carbonates, evaporites, and red beds occur in the Guadalupian (Upper Permian) San Andres Formation (Fig. 45) at many locations on the Northwest Shelf of the Delaware Basin and the eastern and western margins of the Central Basin Platform in West Texas (Fig. 46). Although individual cycles are thin (typically <50 ft thick), reservoirs are laterally extensive (up to 10's of miles wide and long) and host enormous quantities of oil. Stacking of multiple cycles results in the large petroleum accumulations of the Permian Basin. At Slaughter-Levelland and Milnesand fields on the Northwest Shelf of the Delaware Basin (Fig. 47), Mg-rich brines produced during gypsum and anhydrite precipitation on sabkhas flowed downward into underlying subtidal sediments and dolomitized them. Oil became stratigraphically trapped in a belt of porous and permeable, dolomitized subtidal-to-intertidal sediments that pinch out to the north within and beneath the evaporitic sabkha facies (Shinn, 1983; Fig. 48). The highest porosity is developed in dolomitized open marine sediments and in tidal channel deposits. Reservoir quality has been further enhanced by leaching of skeletal grains. The supratidal facies is plugged by evaporites and is nonproductive. Cumulative production from the San Andres in the Permian Basin was 2.78 BBO as of 1986, with 1.1 BBO produced from Slaughter-Levelland Field alone (Ward, et al., 1986).

Ruppel (1986), Ruppel and Cander (1988), and Leary and Vogt (1986) have studied San Andres stacked reservoir cycles in several fields on the Central Basin Platform. The entire subtidal to supratidal sequence has been dolomitized in most fields. Dolomitized subtidal intervals are the best producers. Dolomite is micro- to very finely crystalline with crystal sizes of 10 to 100 µm. A cross section through Emma Field on the Central Basin Platform shows the relationship between lithofacies and porosity (Fig. 49).

Leary and Vogt (1986) used geochemistry to identify Late Permian (Guadalupian) brines as the dolomitizing fluid. Brines generated on sabkha surfaces dolomitized supratidal sediments and then refluxed downward to dolomitize underlying intertidal and subtidal sediments. Dolomite that formed in the supratidal environment has an average $\delta^{18}O$ value of +4.5 ‰ PDB (Fig. 50), indicating dolomitization by a brine that was enriched in ^{18}O due to evaporation. Strontium isotope ratios for the dolomite ranges from .7068 to .7071, indicating the brine formed by evaporation of Late Permian seawater (Fig. 51). Intertidal dolomite has an average $\delta^{18}O$ value of +3.6 ‰, consistent with a similar origin. For the subtidal dolomite, Leary and Vogt (1986) recognized two textures: unaltered (fabric-preserving) and altered (fabric-destructive). Fabric-preserving dolomite has an average $\delta^{18}O$ value of +4.7 ‰, also indicating dolomitization by evaporite brine. Fabric-destructive dolomite has an average $\delta^{18}O$ value that is 1.6 ‰ lighter (Fig. 52), suggesting later recrystallization and fabric destruction either at higher temperature or in an isotopically lighter, possibly meteoric, fluid. In both cases, $\delta^{13}C$ displays the heavy values so typical of Late Permian limestones worldwide, indicating that dolomites of all facies have simply inherited their $\delta^{13}C$ signatures from precursor lime sediments.

By using vertical trends in trace element content (e.g., Fe), and Sr isotope ratios, Ruppel and Cander (1988) demonstrated that dolomitizing brines refluxed downward from the siliciclastic-rich supratidal facies to dolomitize the underlying subtidal facies at Emma Field (Fig. 53). The source of Fe and

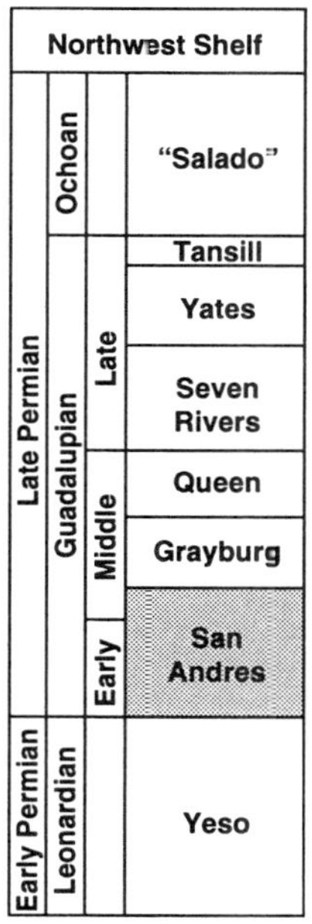

Figure 45. Age and stratigraphic context of San Andres Formation, Permian Basin, West Texas.

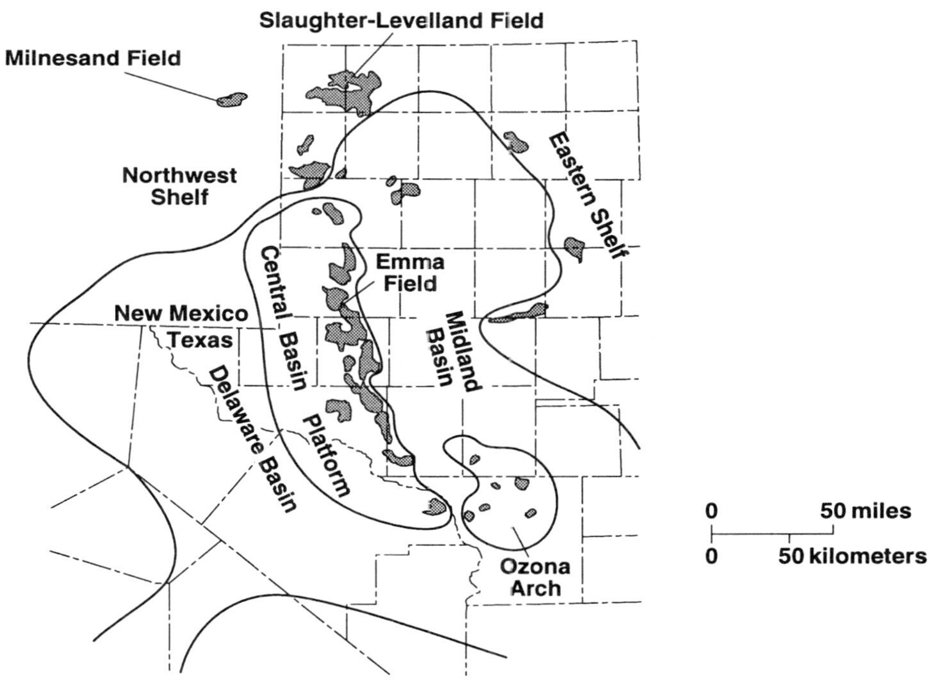

Modified after Hild (1986)

Figure 46. Regional map showing locations of fields that produce from the San Andres Formation around the rim of the Delaware and Midland basins and along the eastern margin of the Central Basin Platform in West Texas.

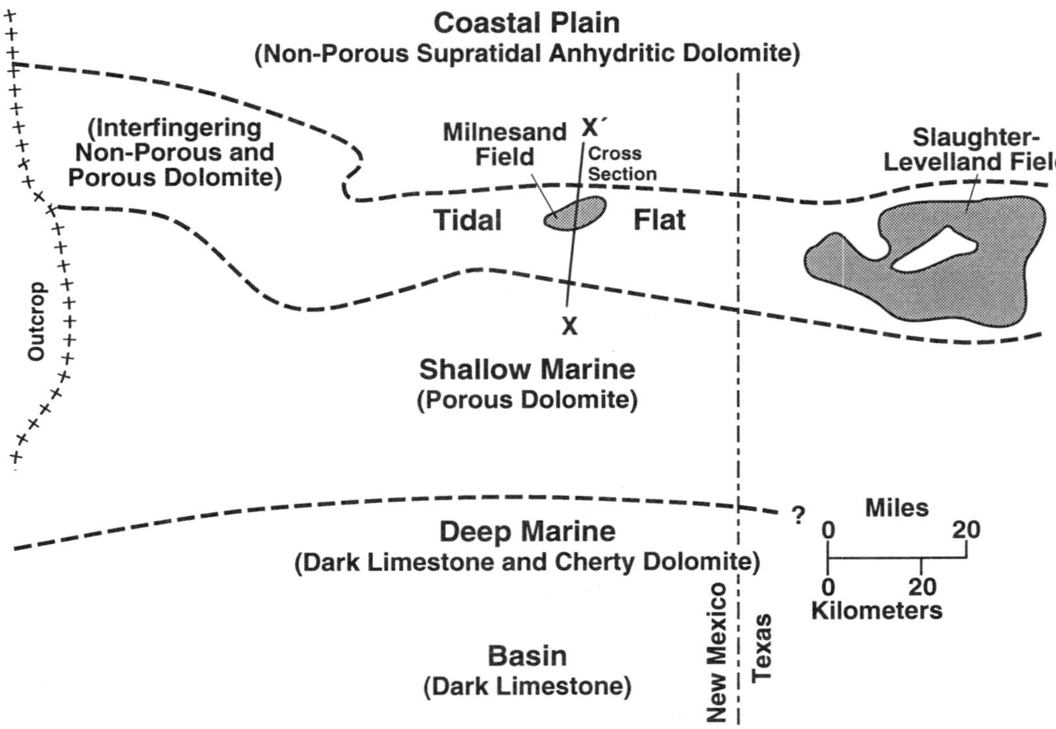

Modified after Shinn (1983)

Figure 47. Location map for Milnesand and Slaughter-Levelland fields, which occur in the San Andres Formation in a tidal flat belt that developed parallel to the northern margin of the Delaware Basin.

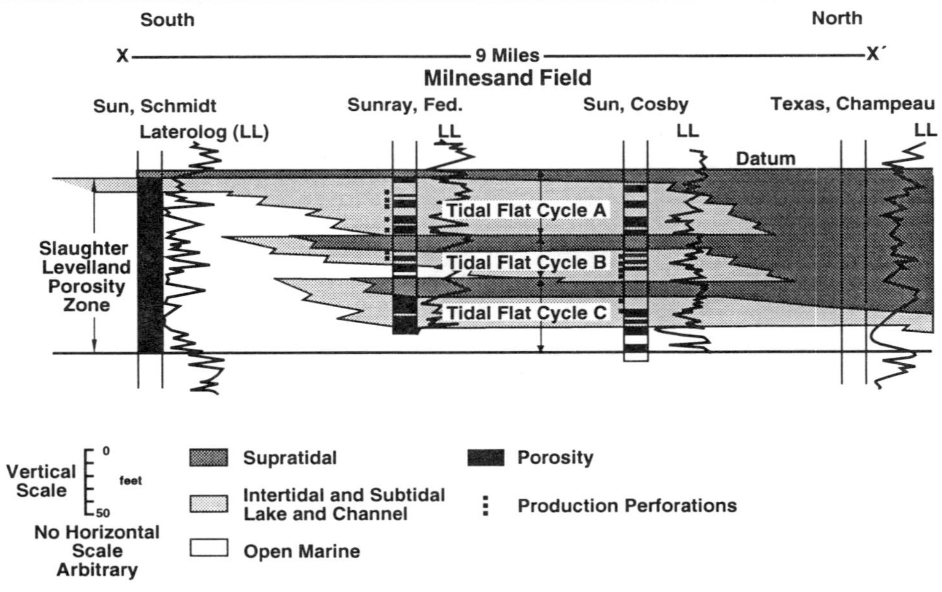

Shinn (1983)

Figure 48. North-south stratigraphic cross section X-X' (Fig. 47) showing three stacked subtidal/intertidal dolomite reservoirs that pinch out to the north and are sealed by impermeable, anhydritic supratidal (sabkha) facies.

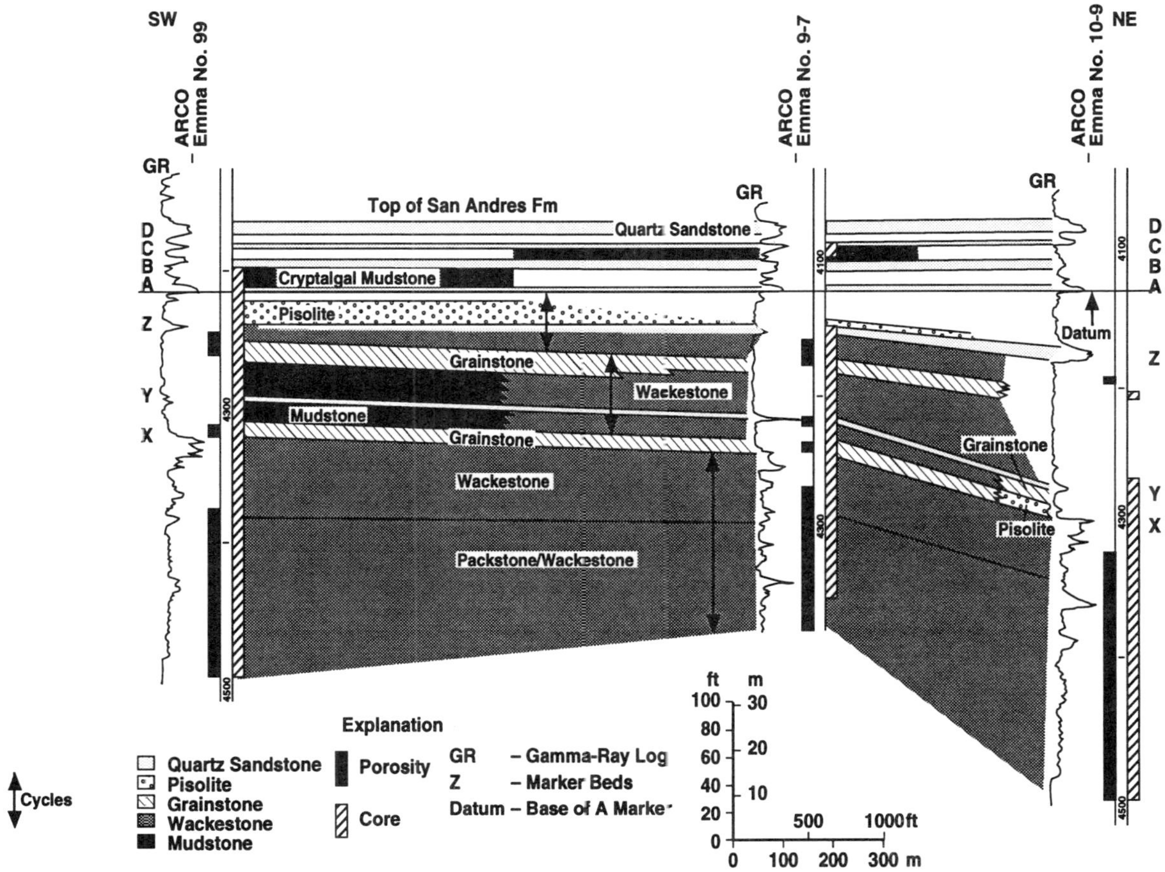

Modified after Ruppel and Cander (1988)

Figure 49. Northeast-southwest cross section through Emma Field on the Central Basin Platform (Fig. 46) showing lithologic distribution within several shoaling-upward cycles. The entire interval has been dolomitized. The cycles are defined by a progression in depositional fabric from mudstone to wackestone to grainstone, culminating in pisolite and quartz sandstone. Note that the best porosity is developed in dolomitized wackestones (subtidal environment).

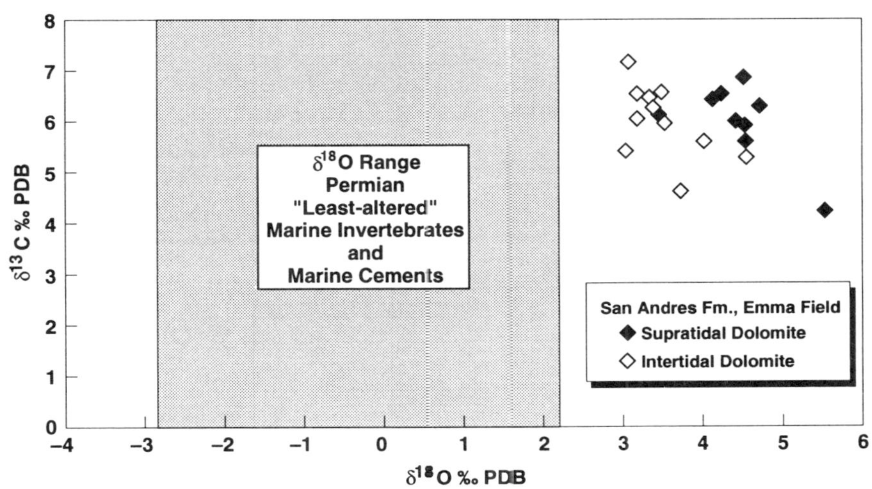

Modified after Leary and Vogt (1986)

Figure 50. Isotopically heavy $\delta^{18}O$ values for San Andres dolomite at Emma Field indicate dolomitization of supratidal and intertidal sediments by an evaporation-concentrated marine brine. $\delta^{13}C$ displays the heavy values typical of Late Permian limestones worldwide, indicating that dolomite has inherited its $\delta^{13}C$ signature from precursor lime sediments.

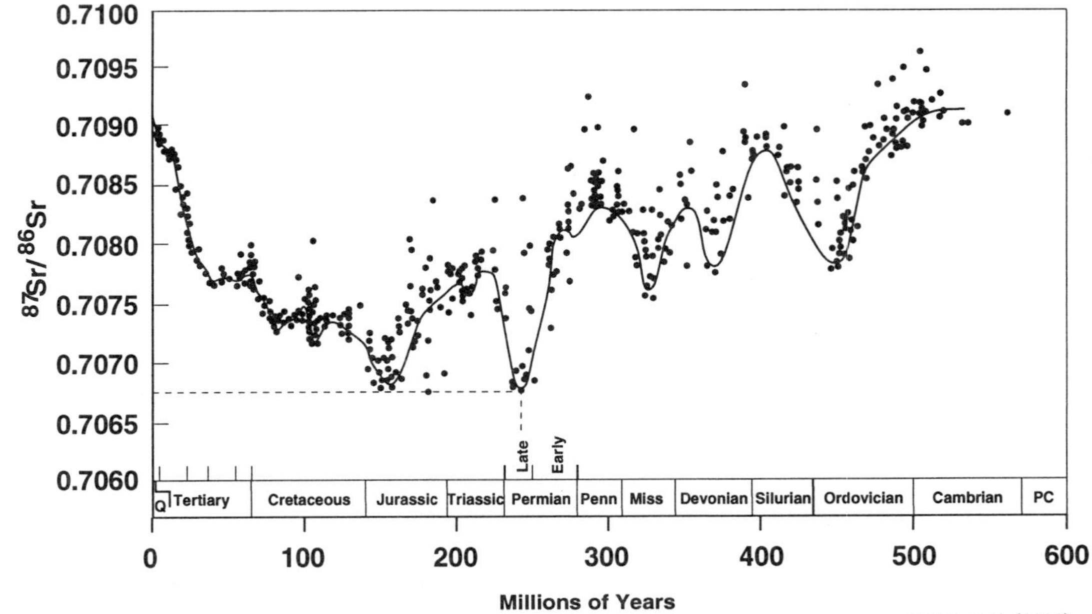

Burke, et al. (1982)

Figure 51. Strontium isotopic composition of seawater plotted as a function of geologic age. Strontium isotope ratio reaches a minimum of .7067 in the Late Permian, imparting a distinctive signature to dolomite that formed from Late Permian seawater-derived brines. The Emma Field dolomite has a Sr isotopic compositional range of.7068 to .7071, which argues strongly for dolomitization by Late Permian brines.

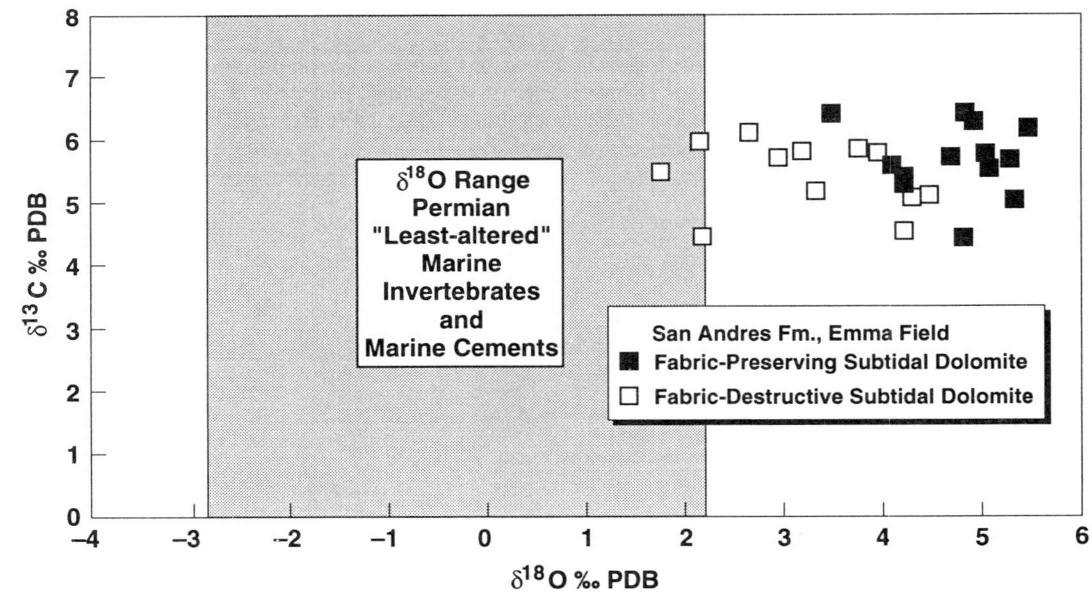

Modified after Leary and Vogt (1986)

Figure 52. Fabric-preserving subtidal dolomite has heavy $\delta^{18}O$ values indicating formation from the same evaporation-concentrated brines that dolomitized supratidal and intertidal sediments (Fig. 50). Fabric-destructive dolomite from the same subtidal facies is depleted in ^{18}O by 1.6‰, suggesting later recrystallization either at higher temperature or in isotopically lighter, possibly meteoric, fluid. Fabric-destructive ("sucrosic") dolomite often forms the best reservoir rock in this setting.

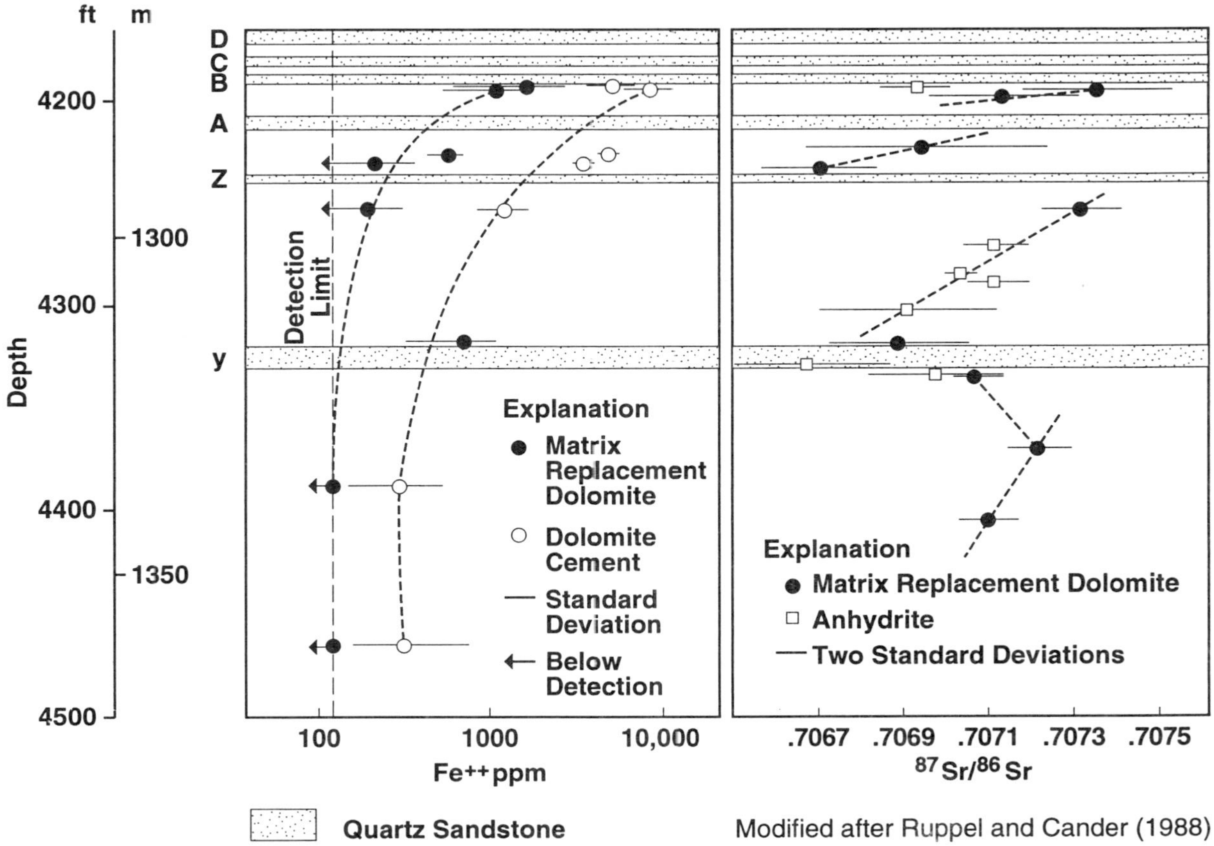

Figure 53. Depth distribution of iron contents (ppm) and Sr isotope ratios in dolomite and anhydrite, San Andres core, Arco Emma No. 99. Iron concentration is greatest in siliciclastic-rich beds in upper San Andres and decreases downsection, indicating both a source of iron in the siliciclastic beds and a downward flow of dolomitizing fluids through the section. Variations in Sr isotope ratio show the same downward-decreasing trend, but display local increases below several siliciclastic beds, indicating each bed served as a source of radiogenic Sr for downward-flowing fluids.

radiogenic Sr (^{87}Sr) in the subtidal dolomite was the siliciclastic interbeds within the supratidal facies of the San Andres Formation and overlying units. They also argued that since several dolomite samples have Sr isotope ratios more similar to Late Guadalupian seawater (.7068) than to Early and Middle Guadalupian seawater (.7071), some of the dolomite must have formed in, or have been overprinted by, Late Guadalupian evaporite brines which flowed downsection into the Early to Middle Guadalupian San Andres Formation. This suggests that San Andres dolomites have a hybrid origin which combines elements of both supratidal and reflux dolomitization.

REFLUX DOLOMITE

Reflux dolomite, as originally defined by Adams and Rhodes (1960), is found where brines, generated in evaporative lagoons or barred basins overlying or updip from limestone units, seep downward through the limestone and dolomitize it. In current practice, any dolomite formed from downward-percolating evaporite brines, including those generated on sabkhas, may be referred to as a reflux dolomite. Evaporation of seawater in hypersaline lagoons or basins causes precipitation of gypsum, which raises the Mg/Ca ratio of the residual evaporite brine. Further evaporation can lead to precipitation of halite and potash evaporites, producing brines with densities of up to 1.30 g/cc. These dense brines displace less dense interstitial fluids as they flow downward, into, and through underlying carbonate sediments and dolomitize them (Fig. 54). Fluid flow generally occurs in high porosity and permeability grainstone aquifers, but dolomitization can further enhance reservoir quality. The overlying evaporites form the seal, often aided by structural closure. Reflux dolomitization can produce reservoirs that range in scale from thin porous units developed beneath small evaporating pans to thick, regional shelf and shelf margin trends. Examples of the larger reflux-dolomite reservoirs include those along the San Andres/Grayburg trend on the Central Basin Platform, Permian Basin (Ward et al., 1986). The Arab-D dolomite in the giant Qatif Field, Saudi Arabia, formed from a reflux brine which originated in the Arab D Anhydrite (Wilson, 1985). Typically, reflux dolomites are sealed by supratidal facies, which may include thick sequences of evaporite.

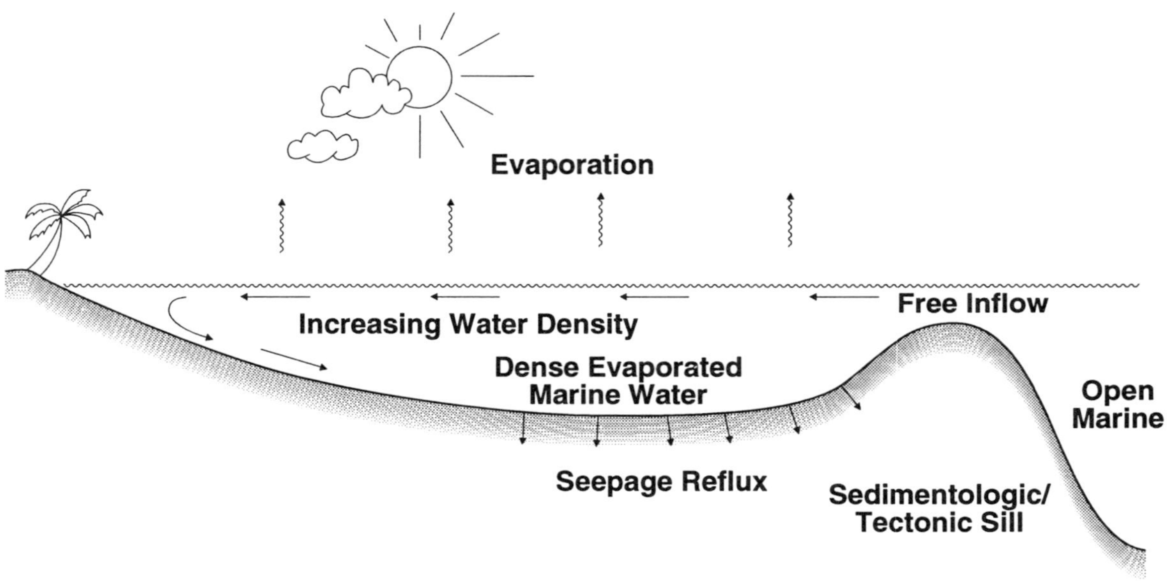

Modified after Moore (1989)

Figure 54. Schematic cross section of a silled evaporitic basin or shelf lagoon. The sill may be either sedimentologic (e.g., a reef or sand shoal), or tectonic (e.g., a horst block). As seawater evaporates, water density increases and gypsum may be precipitated, increasing the Mg/Ca ratio of the brine and enhancing its ability to dolomitize. Dense evaporitic brines sink downward, reflux through the basin or lagoon floor, and dolomitize any carbonate sediments with which they come into contact.

RECOGNITION

Reflux dolomite can be recognized in a variety of ways. It generally occurs downsection or downdip from evaporites. Percent dolomite decreases downward, away from the evaporites. The regional distribution of the dolomite often mimics that of the overlying evaporite unit and crosscuts depositional contacts. Dolomite commonly appears to have formed prior to compaction and is often fabric-preserving. Dolomite crystal size can vary widely from micro- to medium crystalline. Because the reflux process occurs prior to compaction, often the dolomite reservoir possesses a well-distributed network of intercrystalline porosity characterized by smooth pore walls and polyhedral shapes (e.g., Mississippian Mission Canyon Fm, Williston Basin; Lindsay and Kendall, 1985).

Although carbon and oxygen isotopes may not always be definitive, because dolomite can later recrystallize during meteoric flushing or burial, reflux dolomite tends to have **heavy $\delta^{18}O$ compositions and contain predominantly single-phase fluid inclusions**, suggesting low temperatures of precipitation. Strontium isotopes are very useful for identifying reflux dolomite, since this dolomite commonly **assumes the Sr isotope ratio of the evaporite from which the refluxing brine originated**. Thus, Sr isotope values fall on the seawater Sr isotope curve (Burke, et al., 1982) at the point corresponding to the age of the *evaporite deposit* from which the dolomitizing brine was derived (Moore, 1989; Warren, 1989).

SMACKOVER FORMATION

The Jurassic Smackover Formation of East Texas (Fig. 55) provides a good example of reflux dolomitization of an open shelf wedge of ooid, pellet, and bioclastic grainstones (Moore, et al.,1988; Fig. 56A). The grainstones are overlain in part by lagoonal and supratidal Buckner Formation evaporites. Reflux dolomitization of underlying Smackover grainstones (Fig. 56B) follows the general trend of the evaporite depocenter (Fig. 56C) and the halite trend within the depocenter (Fig. 56D). Residual Buckner brines lost their potential to dolomitize as they refluxed downward and achieved equilibrium with Smackover calcite and dolomite. Figure 57 illustrates how this results in a dolomitization gradient in Smackover grainstones, from 100% dolomite at the contact with the Buckner evaporites to <10% dolomite 50 ft downcore. This dolomite forms the major petroleum reservoir facies in the Smackover in East Texas. Pervasive dolomitization extends 50 to 100 ft below the Buckner contact and is continuous throughout an area totalling >3000 km^2 (Moore, et al., 1988). Porosity is both oomoldic and intercrystalline. The highest porosities and permeabilities generally occur in the leached and dolomitized portions of Smackover ooid grainstones ("oomoldic dolomites"). Closure was created by normal faulting and tilting of interfault blocks, salt anticlines, and basement structures (Collins, 1980; Harwood and Fontana, 1984). The top seal was provided by overlying Buckner evaporites. Most Smackover reservoirs in East Texas occur at 10,000 to 13,000 ft depth and contain sour wet gas. About two dozen fields have been discovered in the trend. The largest have recoverable reserves of up to several hundred BCF yielding >10 MMBbls liquids (Collins, 1980).

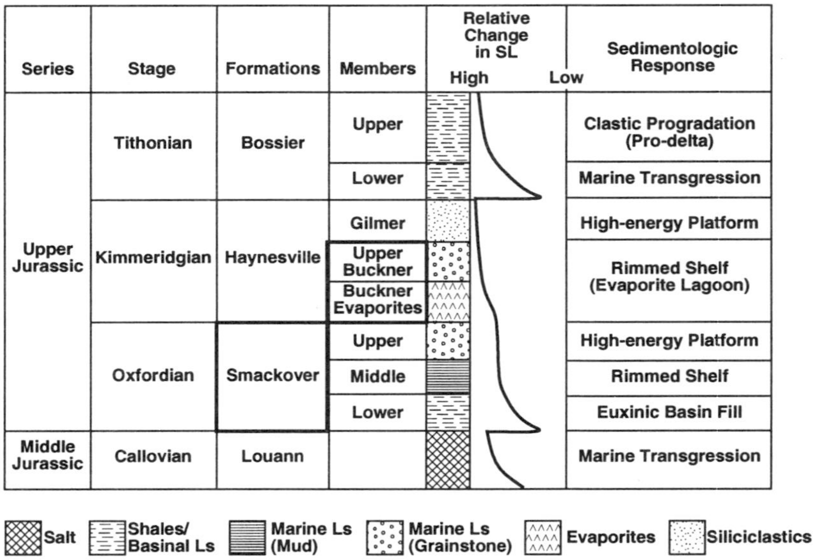

Moore, et al. (1988)

Figure 55. Stratigraphic column and sedimentologic response to sea level change, Upper Jurassic, Gulf of Mexico.

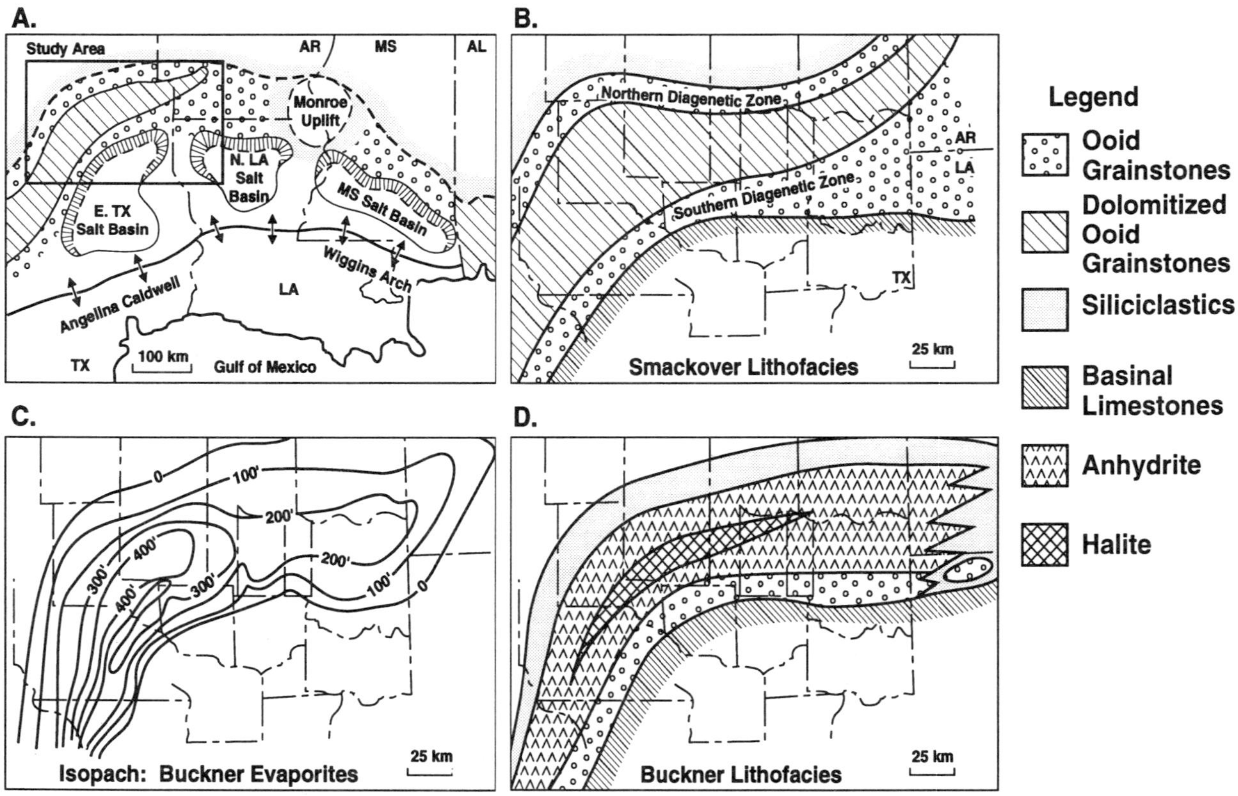

Modified after Moore, et al. (1988)

Figure 56. (A) Structural features controlling Smackover deposition. Areas of oolite deposition, dolomitization, and siliciclastic influx on shelves and deeper water limestone deposition in intrashelf basins are shown. Area examined in Figs. 56B, C, and D is labelled "Study Area" and outlined in black. **(B)** Distribution of upper Smackover ooid grainstones and pervasive dolomite, eastern Texas and southwestern Arkansas. **(C)** Isopach map, Buckner Evaporite, Haynesville Formation. Note that pervasive dolomite in Smackover (Fig. 56B) occurs beneath thickest development of Buckner evaporites (this figure). **(D)** Lithofacies map, Buckner Evaporite. Pervasive dolomite in Smackover (Fig. 56B) is centered beneath halite trend in Buckner (this figure).

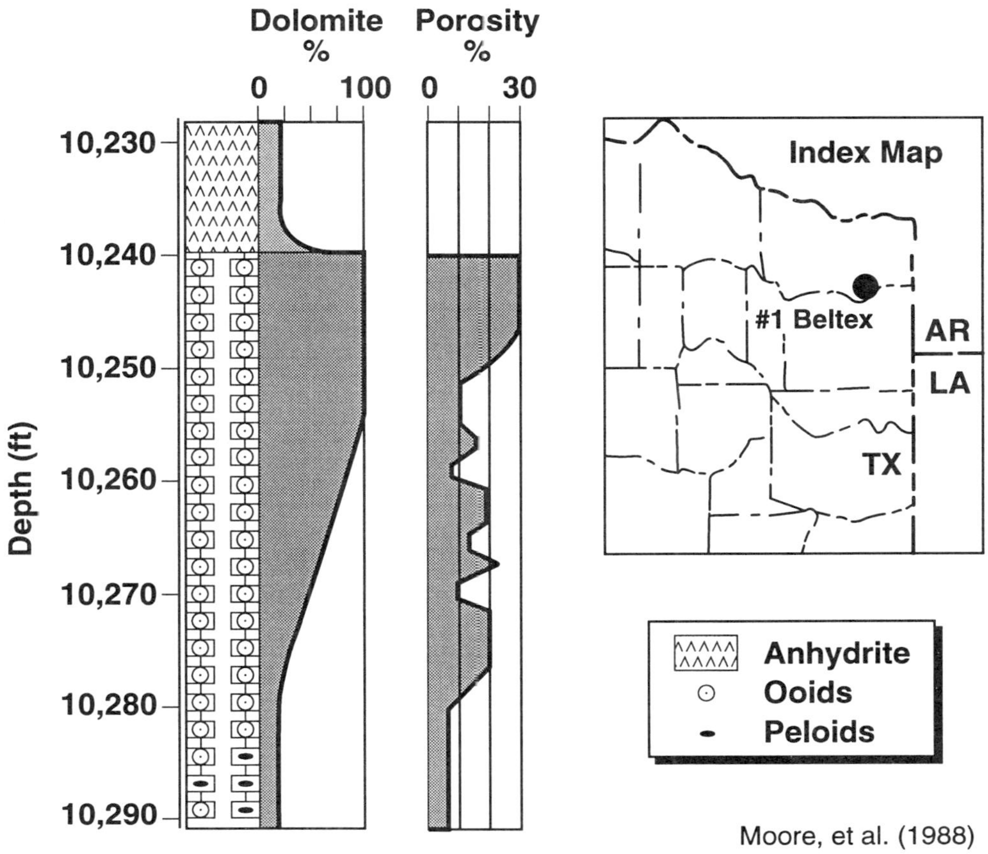

Figure 57. Vertical gradient in dolomitization of Smackover ooid grainstones, from 100% dolomite at contact with Buckner evaporites to <10% dolomite 50 ft downcore. Humble No. 1 Beltex Bowie County, Texas.

Dolomite containing oomoldic porosity is the primary reservoir lithology. Dolomite formed prior to compaction. Crystal sizes in the dolomite matrix between the oomolds range from 5 to 200 µm. Finer-sized crystals occur toward the platform interior, and coarser crystals toward the platform margin (Fig. 58A). Coarsening is caused by growth of clear rims on cloudy crystal cores toward the platform margin. Two additional dolomite phases, of probable burial origin, which are detrimental to reservoir quality, are: (1) post-compaction, euhedral, zoned, replacement dolomite associated with stylolites, and (2) late saddle dolomite cement. These two dolomite phases tend to be depleted in ^{18}O and enriched in Fe, reflecting their burial origins. They are minor constituents and will not be discussed here.

Figure 58B is a $\delta^{18}O$ vs $\delta^{13}C$ crossplot which compares the isotopic compositions of Smackover oomoldic dolomite to the isotopic compositions of evaporitic dolomite from the overlying Buckner Formation. Although both have relatively heavy $\delta^{18}O$ values, indicating early origins, Smackover oomoldic dolomite is depleted in ^{18}O compared to Buckner dolomite. Oomoldic dolomite is also significantly depleted in Sr and Na compared to sabkha dolomite (Fig. 58C). Oomoldic dolomite contains only single-phase fluid inclusions. Thus, although isotopes and fluid inclusions support early dolomitization, and vertical patterns of dolomite distribution indicate stabilization by down-

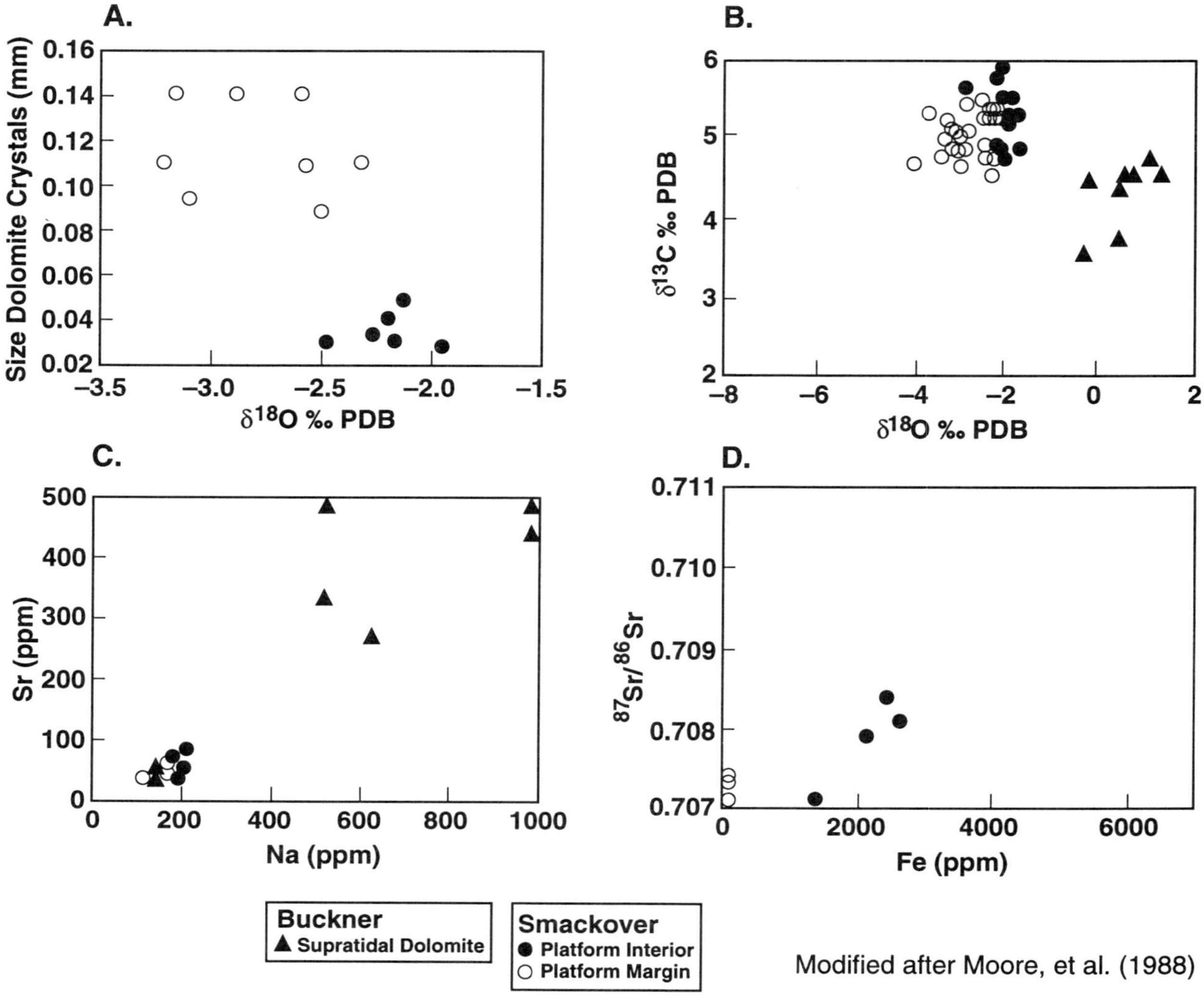

Modified after Moore, et al. (1988)

Figure 58. **(A)** Platform interior dolomite is extremely fine crystalline. Dolomite coarsens to very fine to fine crystalline toward platform margin. Coarsening is accomplished by growth of clear rims on cloudy crystal cores and accompanied by shift to more negative $\delta^{18}O$ values. **(B)** $\delta^{18}O$ vs $\delta^{13}C$ crossplot for Smackover oomoldic dolomite and Buckner supratidal and evaporitic lagoon dolomite. Although both have relatively heavy $\delta^{18}O$ values indicating early origins, Smackover dolomite is depleted in ^{18}O compared to Buckner dolomite, suggesting later recrystallization of oomoldic dolomite in meteoric or burial waters. **(C)** Sr vs Na crossplot for Smackover oomoldic dolomite and Buckner supratidal and evaporitic lagoon dolomite. Smackover dolomite is depleted in Sr and Na compared to Buckner dolomite, also supporting later recrystallization of oomoldic dolomite in meteoric or burial waters. **(D)** Platform interior dolomite is enriched in ^{87}Sr and Fe compared to platform margin dolomite. This, along with the depleted Sr and Na contents of the oomoldic dolomite, suggests that meteoric fluids entered the Smackover at its updip limit where it interfingers with siliciclastic sediments and flowed basinward, recrystallizing and overprinting pre-existing dolomite.

ward refluxing Buckner brines, depleted trace element contents and $\delta^{18}O$ compositions suggest later recrystallization of oomoldic dolomite in meteoric or burial waters.

Regional geochemical gradients from platform interior to platform margin facies appear to point out the direction of flow for fluids responsible for recrystallization. Platform interior oomoldic dolomite is enriched in ^{87}Sr, Fe, and ^{18}O compared to platform margin oomoldic dolomite (Figs. 58B and 58D). The Smackover interfingers with siliciclastic sediments at its updip limit. The elevated ^{87}Sr and Fe contents of platform interior dolomite probably reflect recrystallization of earlier-formed reflux dolomite in meteoric waters that recharged through the siliciclastics. The decrease in $\delta^{18}O$ and increase in thickness of clear rims on dolomite crystals toward the basin margin are thought to reflect an increasing overprint of burial diagenesis in a basinward direction.

NIAGARA FORMATION

Carbonates of the Silurian Niagara Formation in northern Michigan (Fig. 59) provide a good example of reflux dolomitization of pinnacle reef reservoirs. Pinnacle reefs are located on a ramp in a zone which lies between a shelf margin reef-bank complex and deeper water facies in the Michigan Basin (Fig. 60). The reefs range from 0.2 to 2.0 km^2 in area and average 0.5 km^2. More than 500 reefs have been discovered (Sears and Lucia, 1980). The pinnacles are composed of corals, stromatoporoids, crinoids, bryozoans, and brachiopods in a micritic matrix (Cercone and Lohmann, 1985). They are surrounded, overlain, and interfingered with carbonates, which represent times of restricted marine deposition (the A-1 and A-2 Carbonates), and evaporites, which represent times of basin desiccation (the A-1 and A-2 Evaporites; Fig. 61). Although a fabric-retentive dolomite attributed to freshwater-seawater mixing is observed in the lower portions of some reefs, fabric-destructive dolomitization of the upper portions of reefs is most important to reservoir development (Fig. 62; Sears and Lucia, 1980; Bay, 1983). Dolomite is generally finely crystalline and subhedral to anhedral in shape. Dolomitization of the upper portions of reefs is most pervasive in shelfward locations where reefs are overlain by evaporitic tidal flat dolomite of the A-1 Carbonate. As the A-1 Carbonate becomes progressively less dolomitized in a basinward direction, so do underlying reefs (Fig. 63). Within each reef, dolomitization is most complete at the top and decreases downward, away from the overlying tidal flat deposits. Both the vertical and lateral distributions of upper reef dolomite suggest dolomitization was accomplished by downward reflux of brines generated on A-1 Carbonate tidal flats. Reservoir quality was further enhanced by subaerial leaching that occurred at times of sea level lowstand.

Oil generated in organic-rich carbonate members of the Late Silurian Salina Formation in the basin center migrated updip and was stratigraphically trapped in individual pinnacle reefs. Overlying evaporite and carbonate units completely encase the reefs and form the seal (Bay, 1983; Fig. 61). Niagaran reef recoverable reserves for the entire Michigan Basin (which includes both the northern and southern basin margins) are estimated at between 300 and 400 MMBO and 3 to 5 TCF of gas (Gill, 1985).

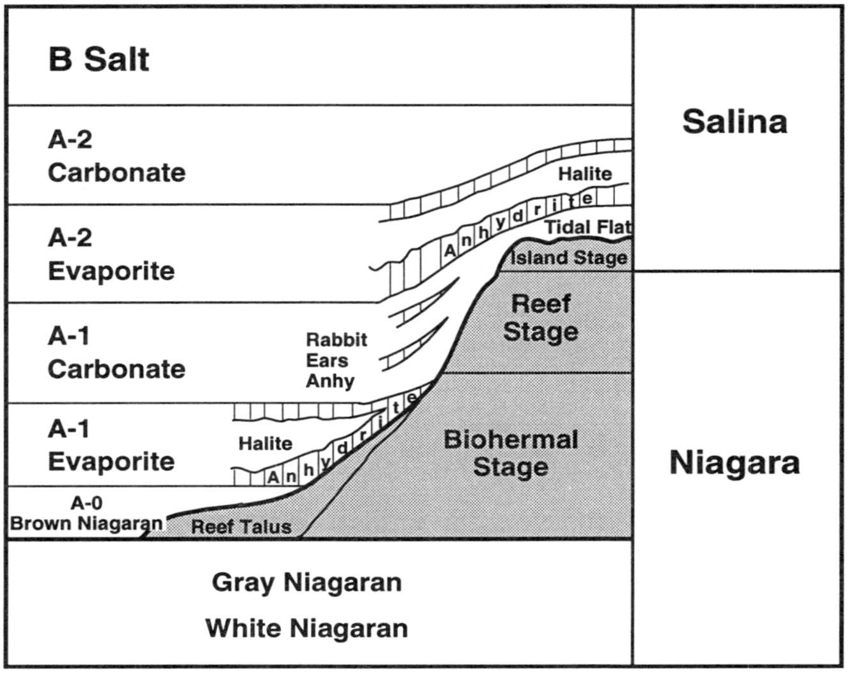

Cercone and Lohman (1985)

Figure 59. Stratigraphic column and relationship of reef to basinal sediments, Niagara and Salina formations, Michigan Basin.

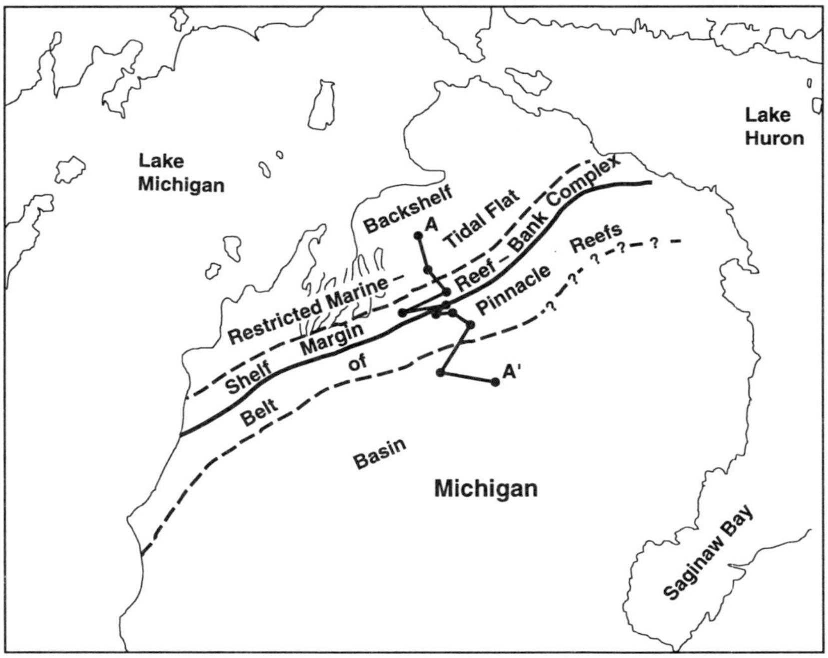

Modified after Bay (1983)

Figure 60. Map of Northern Michigan showing relationship of shelf margin reef-bank complex to pinnacle reef and tidal flat belts. Pinnacle reefs are located in a zone which lies between a shelf edge reef complex and deeper water facies.

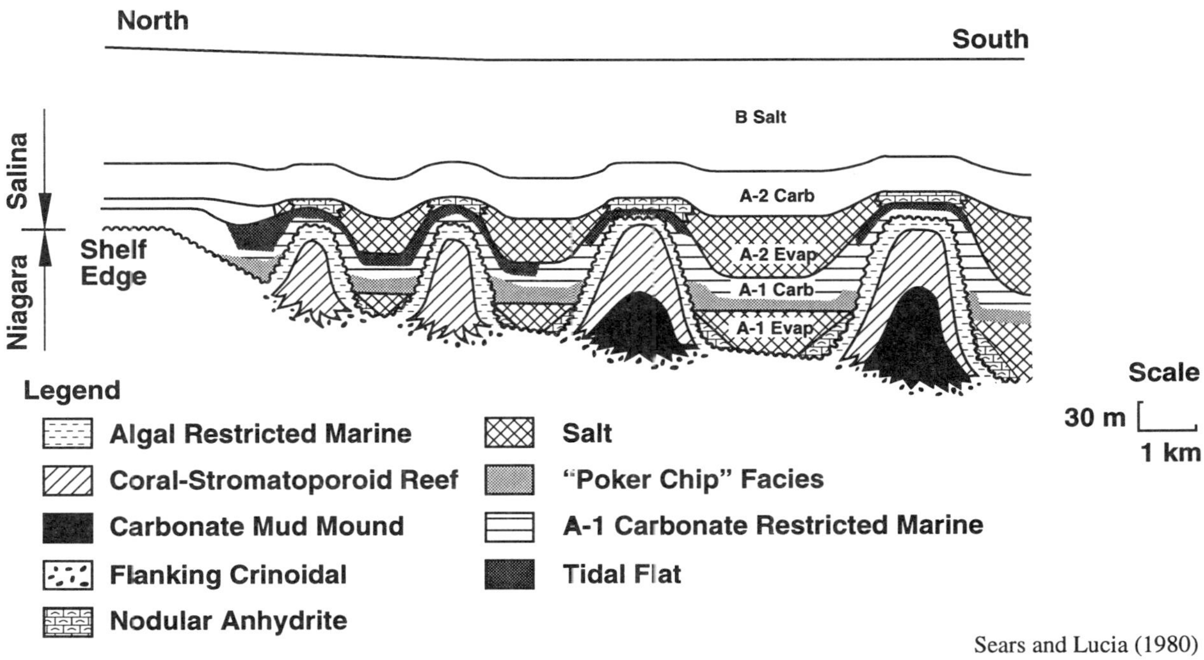

Figure 61. North-South cross section showing facies relationships within reefs and between reef sediments and interreef carbonate/evaporite cycles.

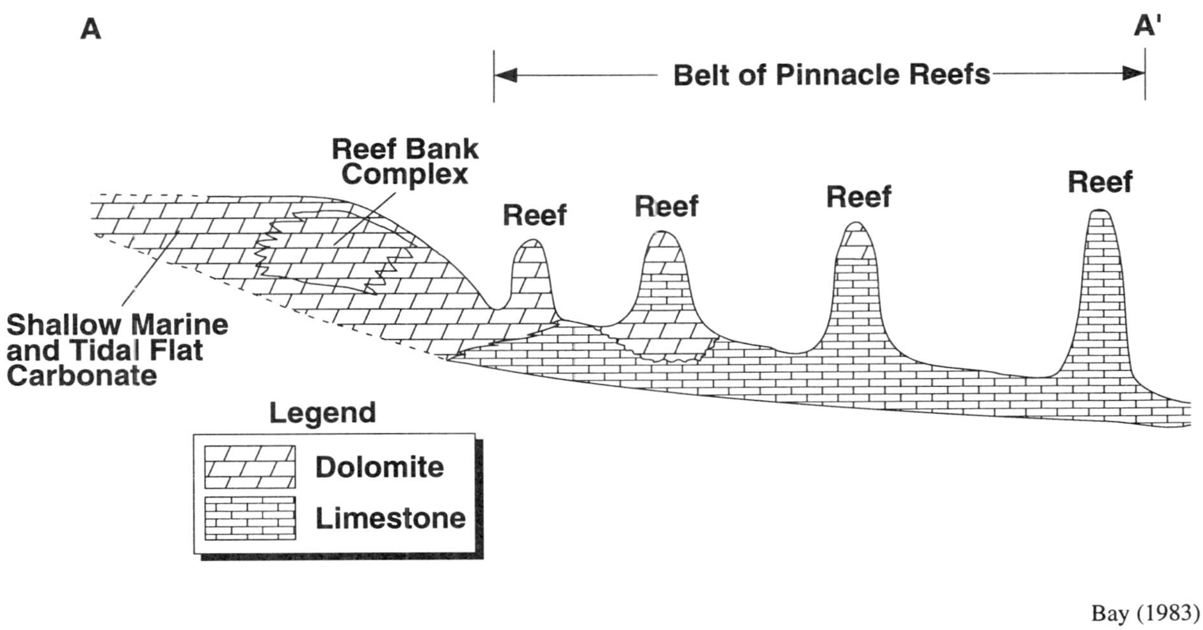

Figure 62. Cross section along line A-A' in Fig. 60 showing dolomite distribution in pinnacle reefs and reef/bank complex. Pervasive fabric-destructive dolomite constitutes highest-quality reservoir and is best-developed in upper portions of pinnacle reefs, beneath evaporitic tidal flats of A-1 Carbonate member. Pinnacles become progressively less dolomitized downsection, suggesting downward reflux of brines. Fabric-retentive dolomite at base of some pinnacles is attributed to freshwater-seawater mixing.

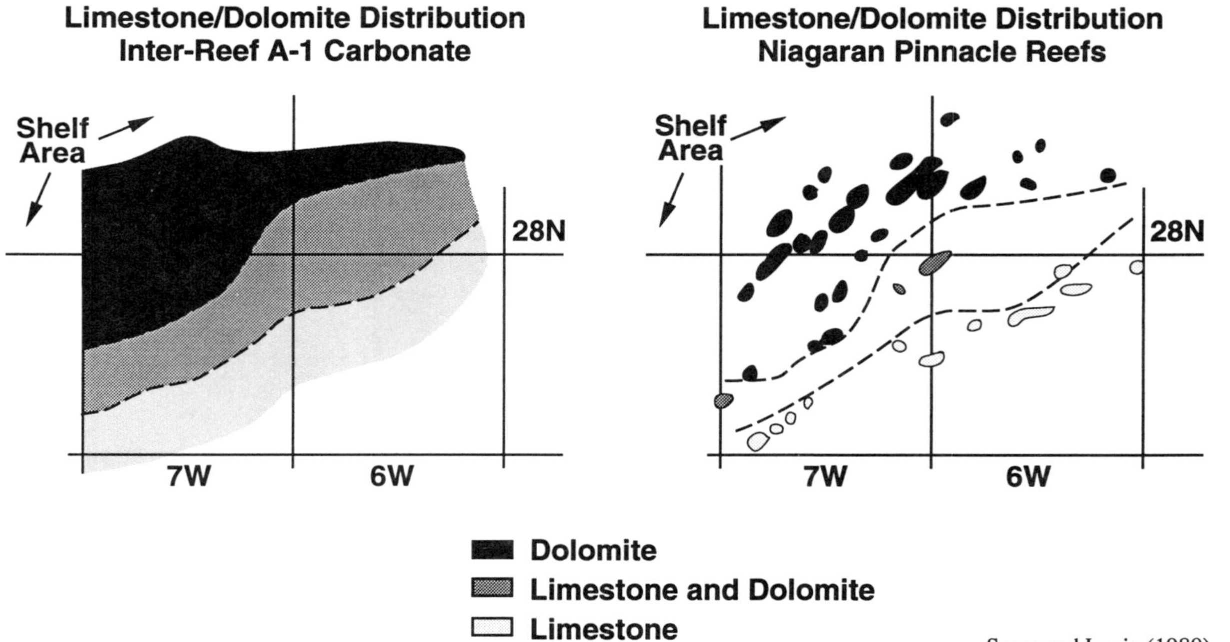

Sears and Lucia (1980)

Figure 63. Dolomitization of pinnacle reefs is most pervasive in shelfward locations where reefs are overlain by evaporitic tidal flats of the A-1 Carbonate. Reefs become less dolomitized basinward, as A-1 Carbonate changes from dolomitized tidal flats to limestone.

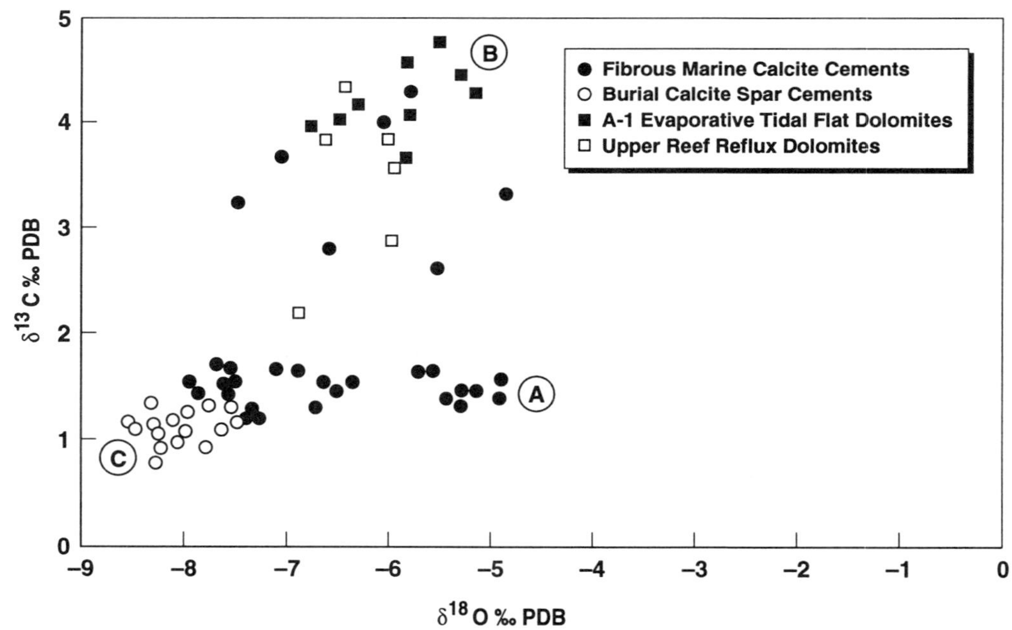

Modified after Cercone and Lohmann (1985, 1987)

Figure 64. $\delta^{18}O$ and $\delta^{13}C$ compositions of fibrous marine calcite cement, burial calcite spar cement, and reflux dolomite from a Niagaran pinnacle reef and evaporitic tidal flat dolomite from the A-1 Carbonate. Isotopic compositions of all phases fall within a region bounded by three isotope endmembers: **(A)** primary marine carbonate precipitated in equilibrium with Silurian seawater; **(B)** A-1 evaporitic tidal flat dolomite; and **(C)** burial calcite cement. Data from Shell 8 State Union well.

Figure 64 shows the $\delta^{13}C$ and $\delta^{18}O$ compositions of fibrous marine calcite cement, burial calcite spar cement, reflux dolomite, and A-1 evaporitic tidal flat dolomite from a cored pinnacle reef (the Shell 8 State Union well; Cercone and Lohmann, 1985, 1987). Marine cement has a fibrous texture and lines pores, which were later filled with marine internal sediment. Burial calcite cement has two-phase fluid inclusions which yield homogenization temperatures of 83 to 108°C. The isotopic compositions of the various phases all fall within a region bounded by three isotope endmembers: (A) primary carbonate precipitated in equilibrium with Silurian seawater (Lohmann and Walker, 1989); (B) A-1 evaporitic tidal flat dolomite; and (C) burial calcite cement.

Burial calcite spar cement has depleted $\delta^{18}O$ compositions that reflect its elevated temperature of precipitation under burial conditions. Marine cement has recrystallized and shows a continuum of $\delta^{18}O$ values from those similar to Silurian marine carbonate to those similar to burial calcite cement. Both marine and burial calcite cements have normal Silurian marine $\delta^{13}C$ values of around +1‰ PDB, whereas A-1 tidal-flat dolomite has enriched $\delta^{13}C$ values of around +4‰ PDB. ^{13}C-enrichment of A-1 dolomite is interpreted to have been caused by incorporation in the dolomite of heavy bicarbonate generated during anaerobic fermentation on organic matter-rich tidal flats. Reflux dolomite in the upper reef has a $\delta^{13}C$ signature intermediate between that of A-1 dolomite and Silurian marine cement. The ^{13}C-enriched isotopic compositions of upper reef dolomite trace the downward flow of tidal-flat brine into the underlying pinnacle reef and provide further evidence for reflux dolomitization (Figs. 64 and 65). Several of the marine cement samples have heavy $\delta^{13}C$ compositions and appear to have recrystallized in the spent dolomitizing solution.

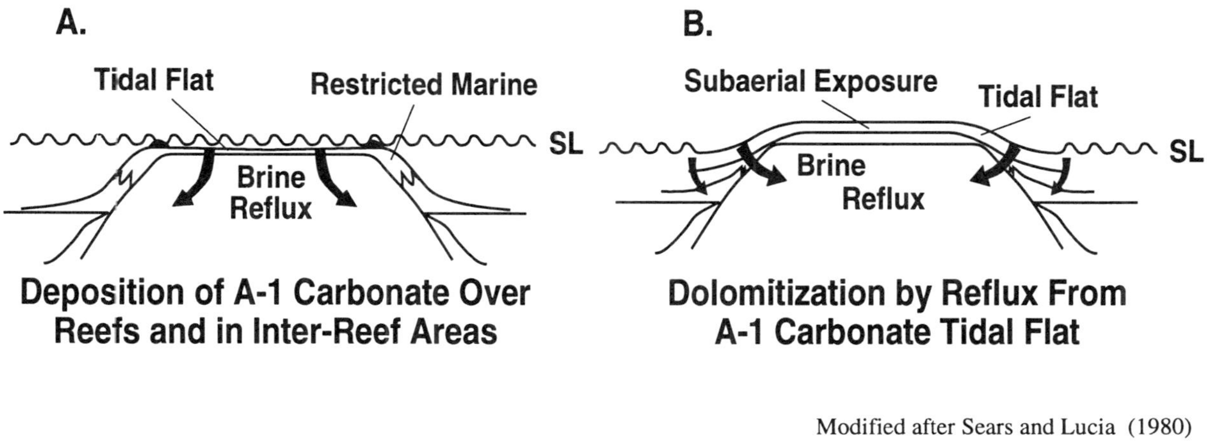

Figure 65. **(A)** During sea level highstand, tidal flats of A-1 Carbonate are deposited above pinnacle reef and brine reflux begins. **(B)** As sea level falls, tidal flat deposition continues, tidal flats prograde beyond reef margin, and reef is dolomitized by downward-refluxing brine.

MARINE AND MIXING ZONE DOLOMITES

Thermodynamic arguments indicate that mixing seawater with fresh water can result in a solution that is undersaturated with respect to calcite and supersaturated with respect to dolomite (Fig. 66). Badiozamani (1973) suggested that dolomitization is favored in mixing zones, which are common in coastal areas where marine interstitial water and fresh water moving seaward in aquifers meet and intermix (Fig. 67A). Over the years, this process has been repeatedly invoked to explain dolomite that is not associated with evaporitic settings or hot subsurface fluids. Many workers have remained skeptical of the mixing model, however, since its thermodynamic foundations have been called into question (e.g., Carpenter, 1976) and few occurrences of dolomite have been documented in modern mixing zones. There is mounting evidence that dolomitization can proceed in normal seawater (Saller, 1984; Carballo, et al., 1987; Mitchell, et al., 1987). Many dolomites that were interpreted as having originated in mixing zones are now being reinterpreted as having formed in marginal marine carbonate sequences through which tidal pumping (Fig. 67B) or Kohout convection (thermally-driven convective half-cells that develop at continental margins; Fig. 67C) have moved large volumes of seawater (Land, 1986, 1989).

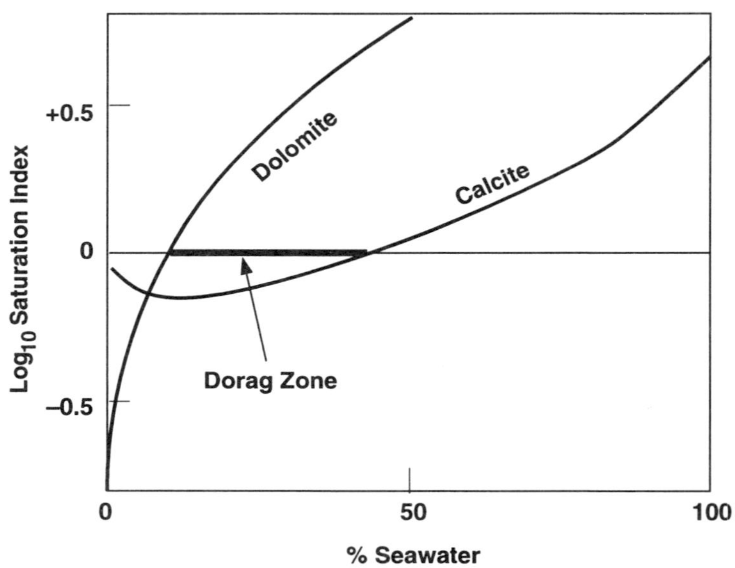

Modified after Hardie (1987)

Figure 66. Thermodynamic basis for marine-meteoric mixing zone dolomitization. When meteoric water, which is saturated with respect to calcite and undersaturated with respect to dolomite, is mixed with seawater, which is supersaturated with respect to both, intermediate mixtures are undersaturated with respect calcite and supersaturated with respect to dolomite. It has been proposed that dolomitization is favored in these intermediate mixtures, which were termed the "Dorag Zone" by Badiozamani (1973).

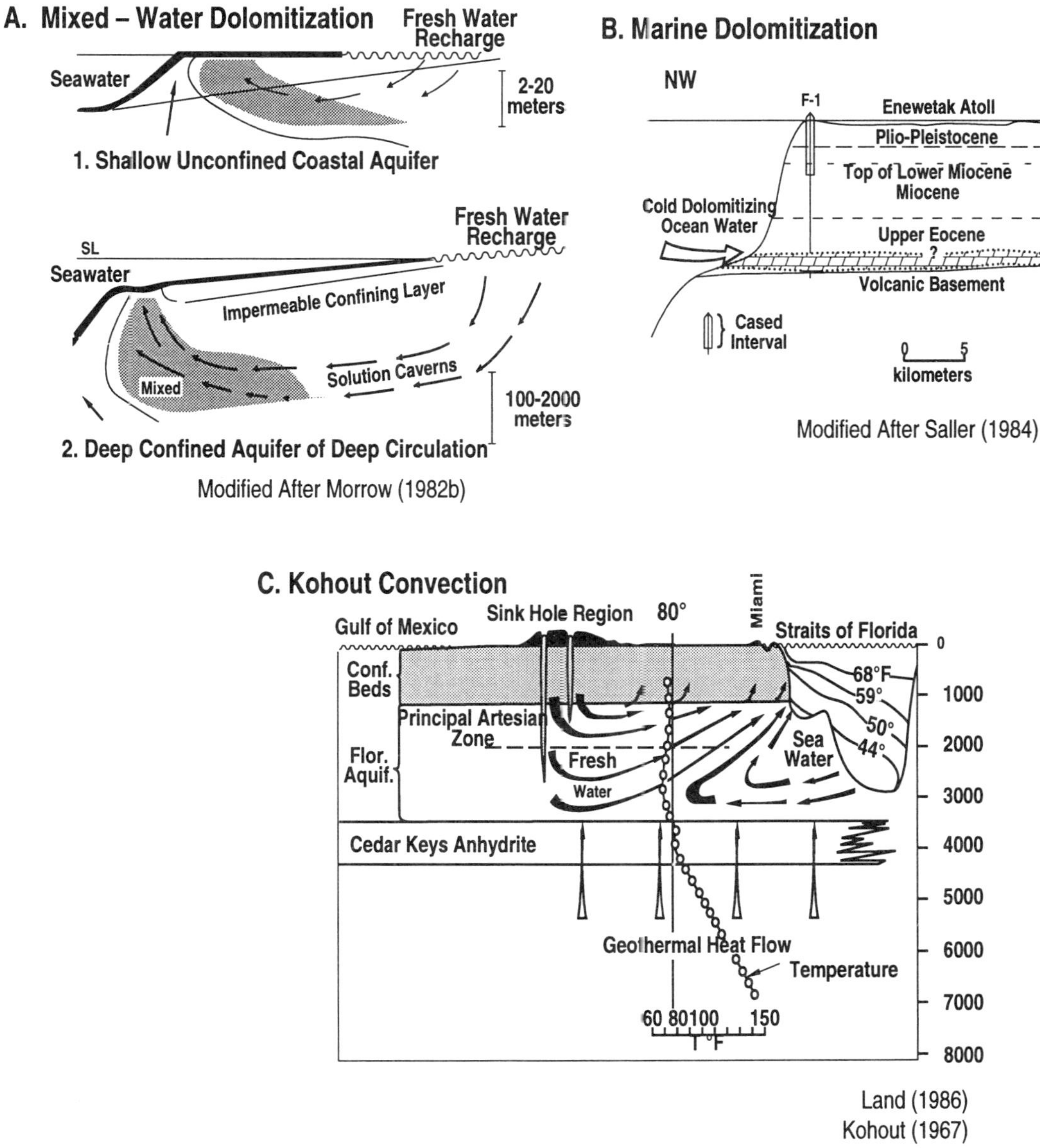

Figure 67. (**A**) Dolomitization in marine-meteoric mixing zones may proceed in either unconfined shallow coastal aquifers (1) or deep confined aquifers (2). Hydrodynamic head forces recharging fresh waters seaward, and the mixing process described in Fig. 66 causes dolomitization. (**B**) Dolomites may form from unmodified seawater. Tidal pumping and thermally driven circulation on Enewetak Atoll caused cold, deep ocean water to circulate through atoll sediments. Carbon, oxygen, and Sr isotope analyses of dolomite in Eocene strata indicate that it formed from Miocene or younger seawater. (**C**) Thermally-driven convective half-cell ("Kohout convection") beneath the Florida Platform. Buoyant rise of geothermally heated water draws in cold seawater at the platform margin. Dolomites may form from seawater alone or from mixing of fresh water and seawater at toe of freshwater aquifer system.

The Chatom Field in Alabama (Smackover Fm) is an example in which mixing zone fluids saturated the crestal part of a dome and resulted in a porous dolomite reservoir (Feasel, 1985). The reservoir is composed of dolomite with intercrystalline and grain-moldic porosity, and is surrounded by non-porous limestone on the flanks of the structure. Dolomites, believed by some to have formed from mixed meteoric-marine fluids, have also been interpreted by others as marine dolomites with subsequent meteoric overprinting. Whatever their origin - marine or mixing zone - these dolomites have several common characteristics. Dolomite bodies are not associated with evaporites and the seal is commonly formed by vertical or lateral facies transition into (or structural juxtaposition against) tight mudstones, wackestones, or shales.

RECOGNITION

Dolomites formed in marginal marine or mixing zone settings have several distinctive characteristics. Perhaps most important are the absence of nearby evaporites and the presence of meteoric diagenetic fabrics, such as moldic porosity and meteoric vadose or phreatic cements. Dolomite cement is usually transparent and relatively free of inclusions, and exhibits complex intercalations of calcite and dolomite, often in the same crystal (Ward and Halley, 1985; Humphrey, 1988). The dolomite is generally finely crystalline, with replacive textures characterized by crystals sizes <20 μm and cement by crystals sizes <100 μm.

Allan and Matthews (1982) suggested that carbonates precipitated in mixing zones between ^{18}O and ^{13}C-enriched seawater and ^{18}O and ^{13}C-depleted fresh water should exhibit positive covariance in their $\delta^{18}O$ and $\delta^{13}C$ values (Fig. 68A). Lohmann (1988b) observed that because mixing zone dolomitization is supposed to occur only in the limited range of seawater-freshwater mixtures that are undersaturated with respect to calcite and supersaturated with respect to dolomite (Fig. 66), mixing zone dolomites may plot parallel to the meteoric calcite $\delta^{18}O$ line, offset toward higher $\delta^{18}O$ values because of the greater dolomite-water fractionation (Fig. 68B). Mixing zone dolomite should contain lower concentrations of Sr and Na than dolomite precipitated from seawater or evaporated seawater. This distinction can be obscured, however, by later recrystallization of marine and evaporitic dolomites in meteoric water. **Marine dolomite should have a Sr isotopic composition similar to that of seawater at the time of sediment deposition, while mixing zone dolomite may contain more radiogenic Sr derived from fresh water which recharged through siliciclastic aquifers.** Because both marine and mixing zone dolomites form in phreatic environments at low temperatures, **both should contain single phase fluid inclusions with low salinities (3.5 wt.% NaCl equivalents for marine dolomite, <3.5 wt.% NaCl equivalents for mixing zone dolomite).**

STE. GENEVIEVE FORMATION

Perhaps the best-documented example of a dolomitized petroleum reservoir developed in a mixing zone setting is the Mississippian Ste. Genevieve Formation at North Bridgeport Field in the Illinois Basin (Choquette and Steinen, 1980; Fig. 69). The North Bridgeport reservoir is developed in oolitic tidal bars, sandy ooid-pellet calcarenite channels, and associated dolomitized lime mudstones and wackestones that are oriented nearly perpendicular to the trend of the La Salle Anticline. Dolo-

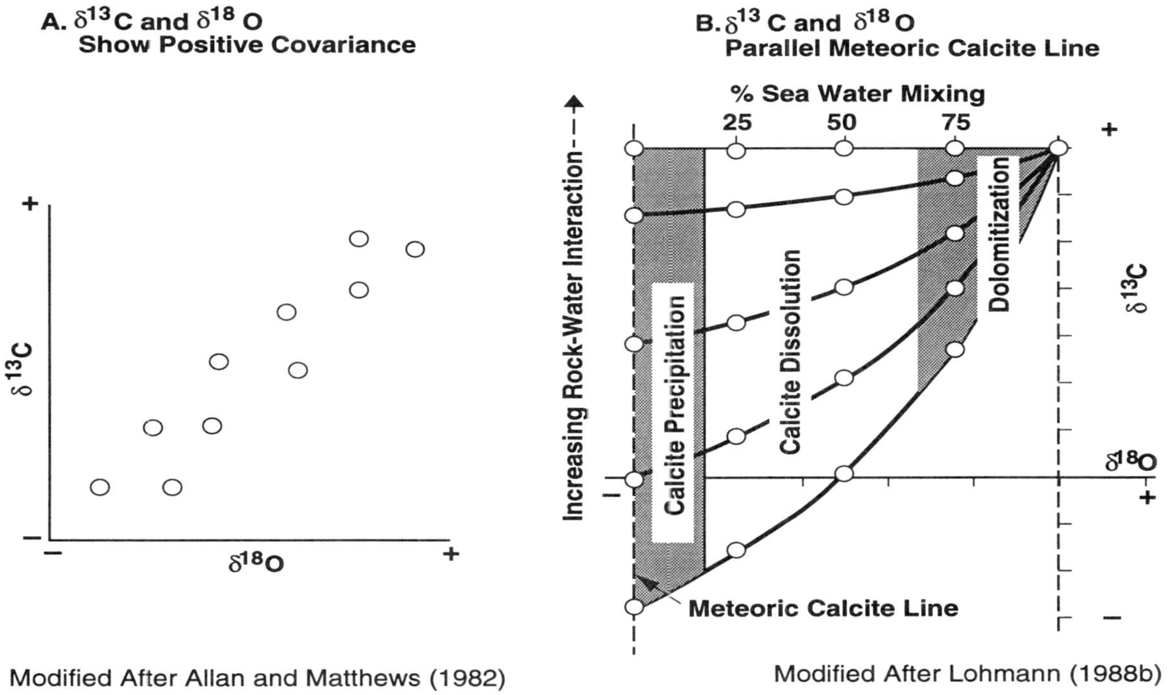

Figure 68. **(A)** Positive covariance of $\delta^{18}O$ and $\delta^{13}C$ in dolomite is caused by mixing ^{18}O- and ^{13}C-enriched seawater with ^{18}O- and ^{13}C-depleted freshwater. **(B)** In actuality, the $\delta^{13}C$ composition of meteoric water can vary considerably, while the $\delta^{13}C$ composition of seawater remains relatively constant. Mixing the two will result in an array of hyperbolic mixing curves which diverge from the seawater endmember composition. Mixing zone dolomites form only in the shaded area to the right, due to the narrow range of seawater/fresh water mixtures that are capable of dolomitization. They may therefore have $\delta^{18}O$ and $\delta^{13}C$ compositions that plot parallel to the meteoric calcite line and show larger variation in $\delta^{13}C$ than in $\delta^{18}O$.

mitized lime mudstones and wackestones generally underlie or are laterally adjacent to elongate lenses of ooid grainstone and sandy ooid-pellet packstone/grainstone (Fig. 70). They have orientations and areal dimensions (0.7 to 4.5 mi long, 0.3 to 1.5 mi wide, and up to 38 ft thick) that are similar to those of the grainstone bodies (Fig. 71). Both the ooid grainstones and the dolomitized wackestone/mudstones contribute to the reservoir, with the grainstones forming a low porosity, high permeability reservoir facies and the wackestone/mudstones forming a high porosity, low permeability reservoir facies. Entrapment is accomplished by structural closure coincident with changes in depositional facies, both laterally and vertically, into low permeability lime mudstones and wackestones (Choquette and Steinen, 1985). Reserves are estimated at 4.0 MMBO recoverable.

Dolomite is microcrystalline, with crystal sizes ranging from 5 to 20 μm. The oolite bodies contain freshwater vadose and beachrock cements, suggesting that episodes of subaerial exposure occurred. Moldic porosity is common in the dolomite and is attributed to meteoric water influence. There is no evidence of supratidal-sabkha conditions or evaporite sequences nearby. The $\delta^{18}O$ and $\delta^{13}C$ compositions and Sr and Na contents of Ste. Genevieve dolomite are very similar to those of Plio-Pleistocene dolomite of Jamaica, Bonaire, and the Yucatan where a mixing zone origin has been

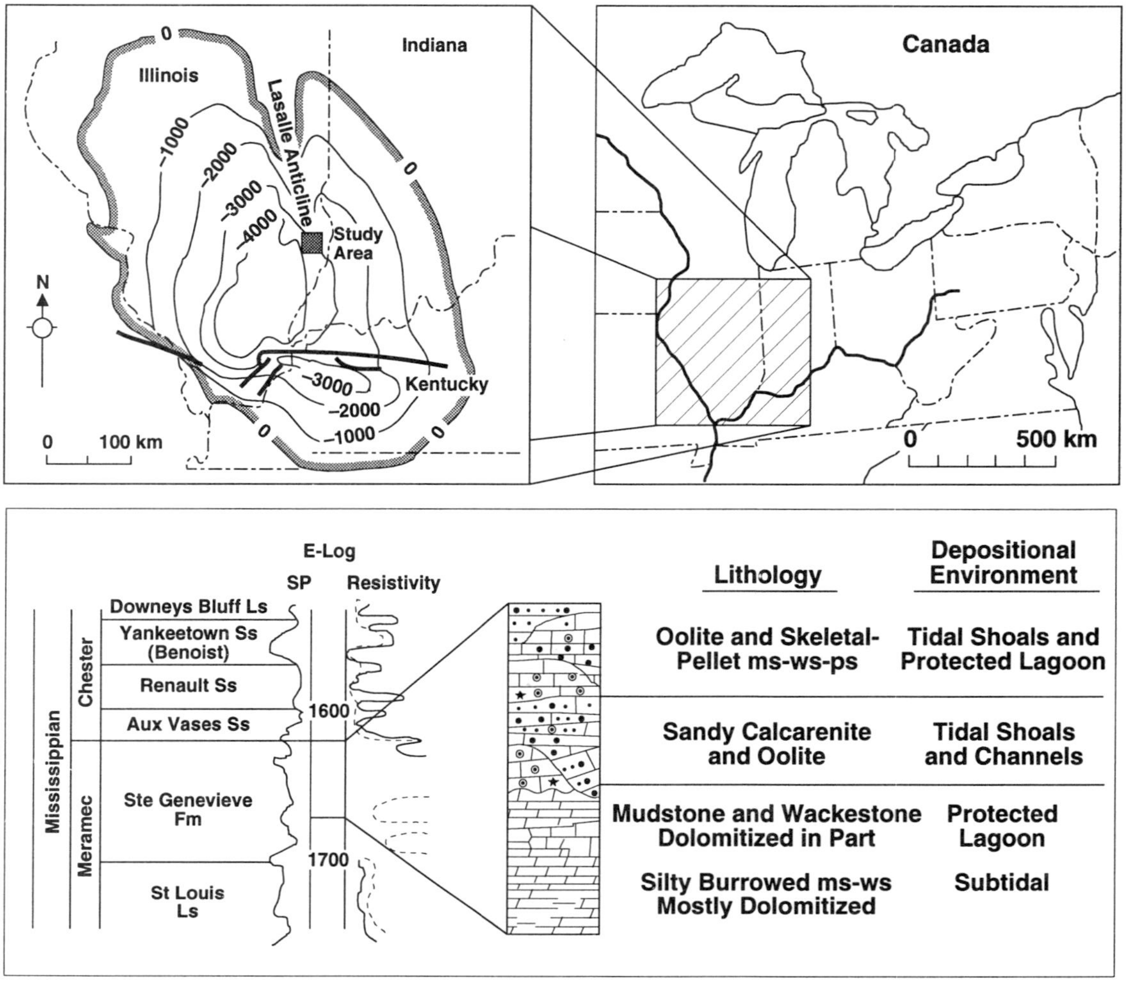

Modified after Choquette and Steinen (1985)

Figure 69. Index map and generalized lithologic column for Ste. Genevieve Formation. Structure is contoured on top of the Upper Devonian-Lower Mississippian New Albany Shale (not shown).

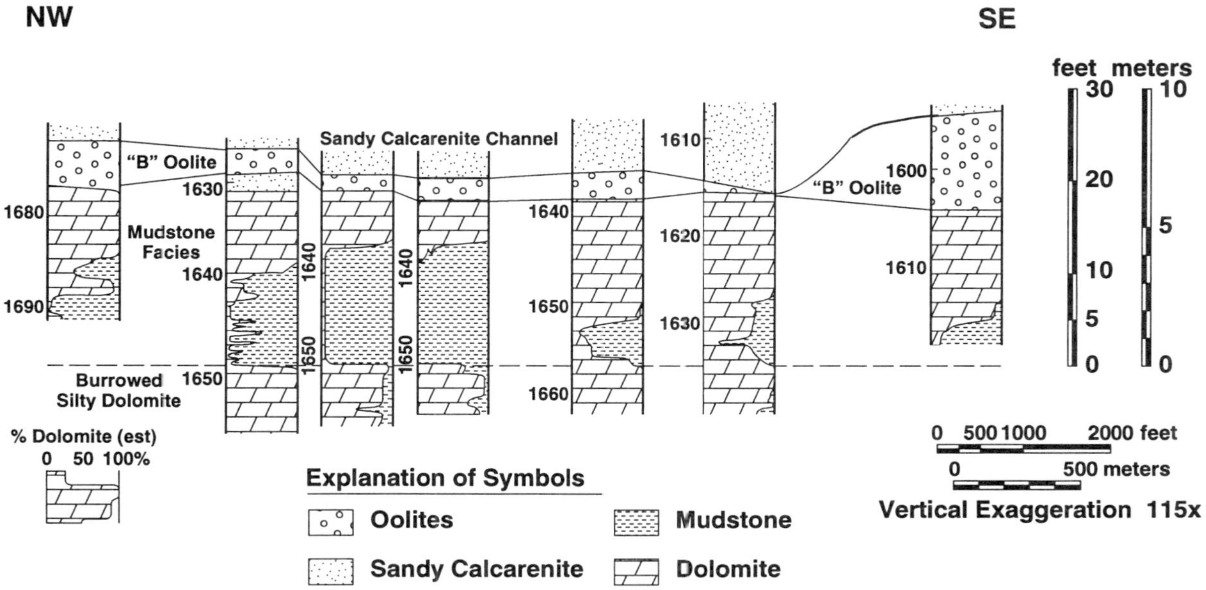

Modified after Choquette and Steinen (1980)

Figure 70. Northwest-southeast cross section, Ste. Genevieve Formation, North Bridgeport Field. Best reservoir development is in oolitic tidal bar limestones and in underlying dolomitized mudstones and wackestones.

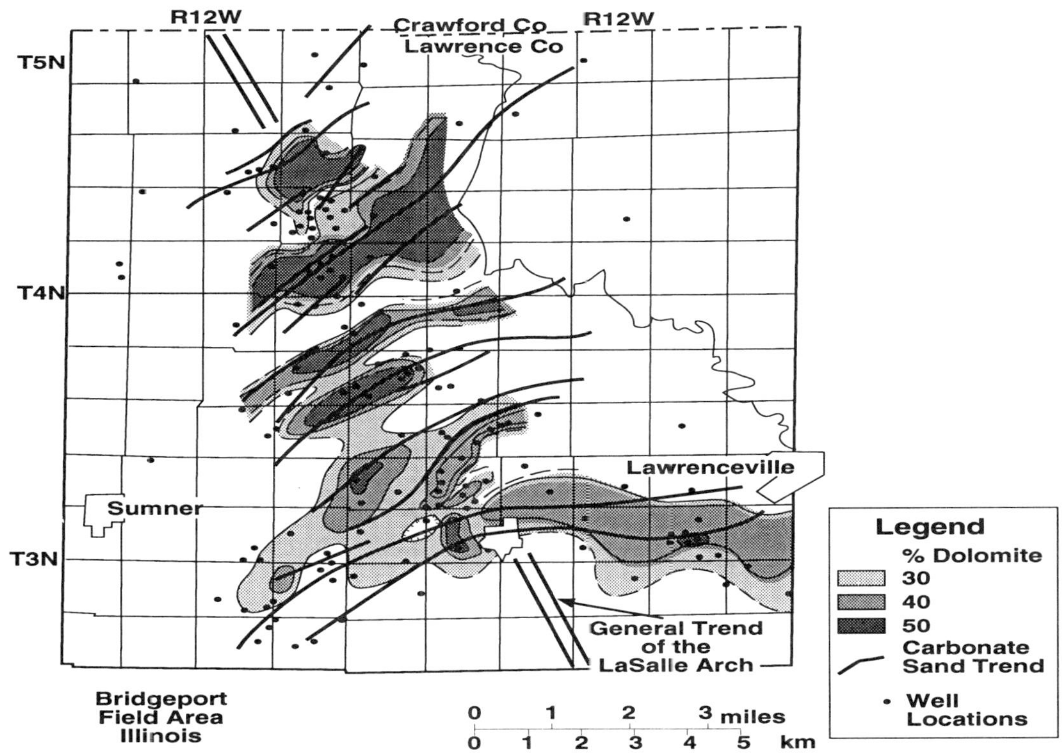

Modified after Choquette and Steinen (1980)

Figure 71. Map of percent dolomite with carbonate sand trends (oolitic tidal bars and sandy calcarenite channels) superimposed. Note the close correspondence between sand trends and dolomitization in the underlying mudstones and wackestones.

invoked (Figs. 72A, 72B and 72C). Therefore, Choquette and Steinen propose that dolomitization is associated with a gravity-driven regional meteoric aquifer, with much of the flow occurring in the porous grainstone channels and shoals. Dolomite formed in the mudstones adjacent to and beneath these permeability conduits where fresh and marine waters mixed (Fig. 73).

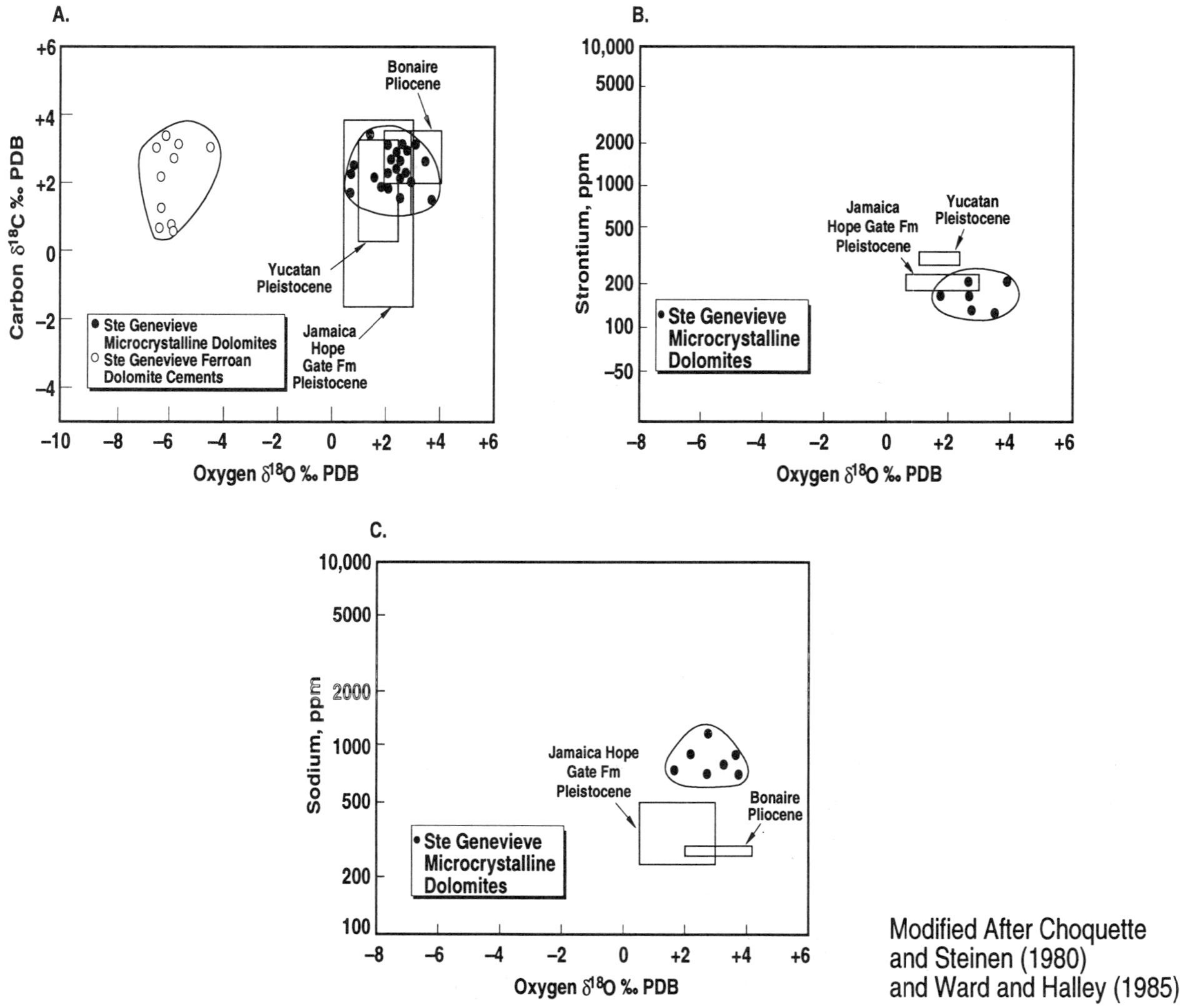

Figure 72. **(A)** $\delta^{18}O$ vs $\delta^{13}C$ values for Ste. Genevieve dolomite compared to values for possible mixing zone dolomites from Jamaica, Bonaire, and the Yucatan. Stable isotopic and trace element compositions of Ste. Genevieve dolomite is almost identical to those of Plio-Pleistocene dolomites of Jamaica, Bonaire, and the Yucatan where a mixing zone origin has been invoked. Close similarities in composition have been used to support mixing zone origin for Ste. Genevieve dolomite. **(B)** $\delta^{18}O$ vs Sr values for Ste. Genevieve dolomite compared to possible mixing zone dolomites from Jamaica and the Yucatan. Note close similarities in composition. **(C)** $\delta^{18}O$ vs Na values for Ste. Genevieve dolomite compared to possible mixing zone dolomites from Jamaica and Bonaire. Note similarities in composition.

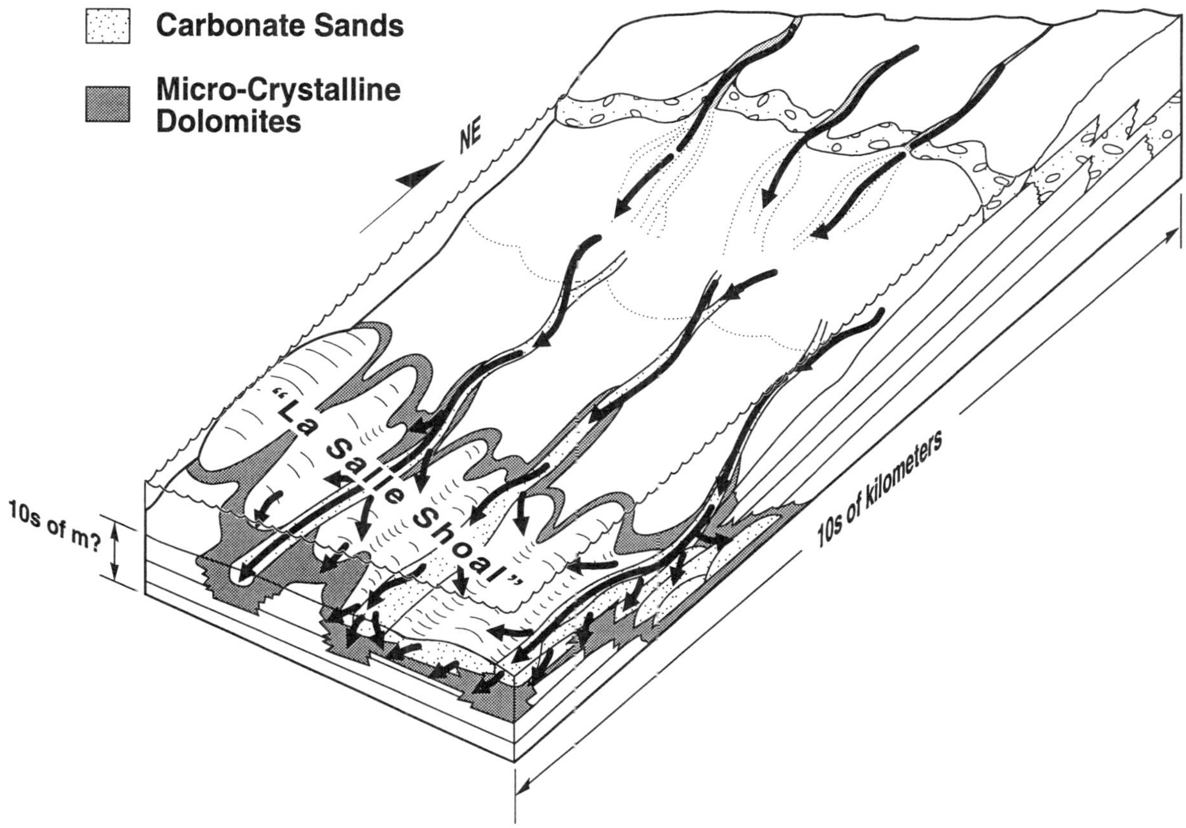

Modified after Choquette and Steinen (1980)

Figure 73. Proposed hydrologic model for dolomitization of Ste. Genevieve. Groundwaters from recharge area to Northeast flowed through carbonate sand aquifers and mixed with seawater beneath sand channels and tidal bars to dolomitize underlying mudstones and wackestones.

BURIAL DOLOMITE

Dolomitization can occur in the subsurface during any part of the burial and uplift history of a carbonate formation. The elevated temperatures encountered during burial favor dolomitization, both by removing kinetic barriers and by raising the Ca/Mg ratio at which dolomitization can occur. Figure 74 shows that above about 60-70°C, most sedimentary formation waters are capable of dolomitizing limestones, whereas at surface temperatures, lower Ca/Mg ratios may be required (Hardie, 1987). As stated earlier, in order to dolomitize a limestone, two requirements must be met: (1) a source of Mg sufficient to dolomitize the limestone must exist; and (2) a transport mechanism must be available to carry the Mg to the site of dolomitization, because mass balance calculations indicate that hundreds of pore volumes of fluid must be exchanged in order to complete the dolomitization reaction. The most-cited fluid sources for burial dolomitization are: (1) Mg-rich residual evaporite brines, (2) modified seawater, and (3) shale compaction waters. The important role of residual evaporite brines in producing burial dolomite has been amply demonstrated by rock and fluid geochemistry and fluid inclusion case studies (Broomhall and Allan, 1987; Carpenter, et al., in prep.; Spencer, 1987). The role of seawater, modified by burial diagenetic processes, is more speculative, and is based largely on the chemical argument that most sedimentary formation waters, which have evolved from seawater, are capable of dolomitizing limestone (e. g., Hardie, 1987). The role of shale compaction waters, which are a mixture of seawater and expelled interlayer water from clay minerals, is most speculative, is based on the observed juxtaposition of some dolomite bodies and shale units, (Illing, 1959; Mattes and Mountjoy, 1980), and suffers from insurmountable mass balance problems (e.g., inadequate quantities of Mg in the fluid) when large volumes of dolomite are involved (Morrow, 1982a).

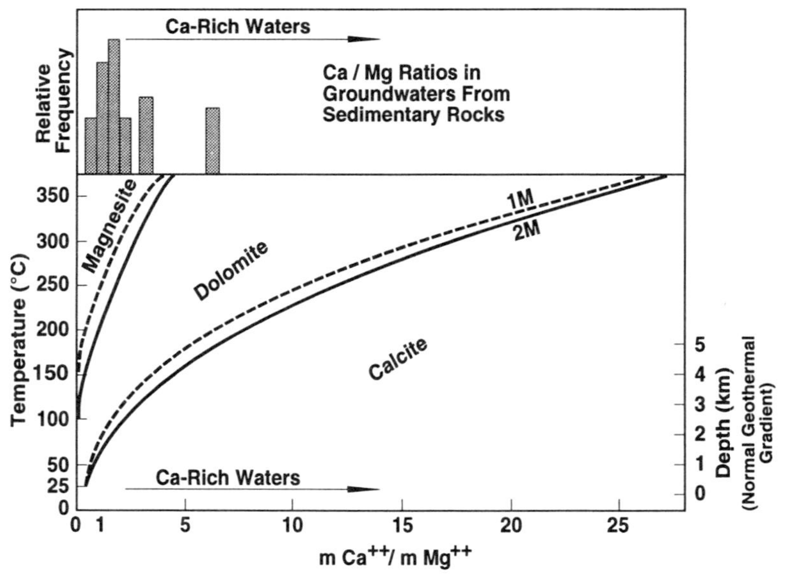

Hardie (1987)

Figure 74. Plot of stability fields of ordered dolomite, calcite and magnesite as a function of temperature and Ca/Mg molar ratio in 1M and 2M chloride brines, extrapolated from high-temperature experimental data. At top is a histogram of Ca/Mg ratios of groundwaters from a variety of sedimentary formations. At temperatures above 60-70°C, most of these groundwaters could be dolomitizing fluids.

Because subsurface fluid flow can be either lateral or cross- formational, burial dolomite reservoirs can assume many shapes and sizes. Hydrodynamic head and elevated density of hypersaline brine both result in lateral and downward flow of formation fluids, while compaction, due to either normal burial or tectonic compression, results in lateral and upward flow. Hydrodynamic head associated with the uplift of mountain belts (Fig. 75; Garven and Freeze, 1984; Bethke, 1986) and compression due to crustal shortening associated with subduction or continental collision (Fig. 76; Oliver, 1986; Leach and Rowan, 1986) are hypotheses for large-scale fluid movement during burial which have gained considerable popularity in recent years. Vertical flow along faults and fractures tends to produce narrow, linear dolomite bodies. An example of this includes the Tangent Field (in the Wabamun Fm, of Alberta, Canada; Stoakes, 1987). Lateral fluid flow can produce thick, tabular dolomite bodies, distributed over hundreds of square miles. This is common where the dolomites result from the dewatering of a large nearby basin. Examples of this include the Berland River-Windfall trend (Woodbend and Winterburn groups) in Alberta. Tabular burial-dolomites can be confined to specific stratigraphic intervals (Broomhall and Allan, 1987) or can ascend and descend through the stratigraphic section on a regional scale (Morrow, et al., 1986). Figures 77, 78, and 79 illustrate three common types of occurrence.

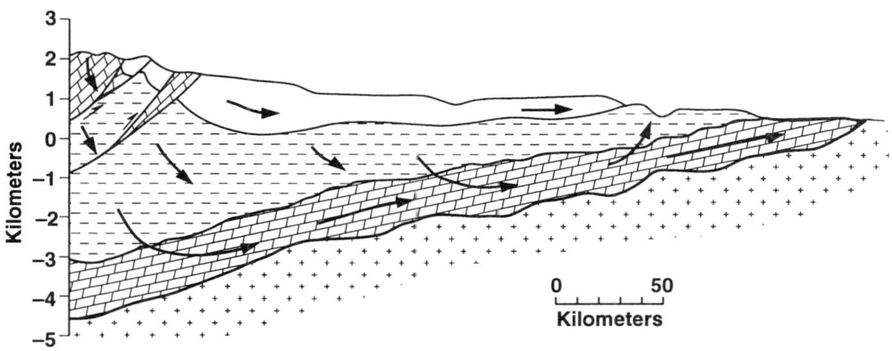

Modified after Garven and Freeze (1984)

Figure 75. Conceptual model of hydrodynamic flow in the subsurface of sedimentary basins. Flow is controlled by topography, water-table configuration, and the subsurface permeability distribution (arrows). Gravity-driven meteoric waters can displace saline formation waters and brines and move dolomitizing solutions laterally and vertically through the basin.

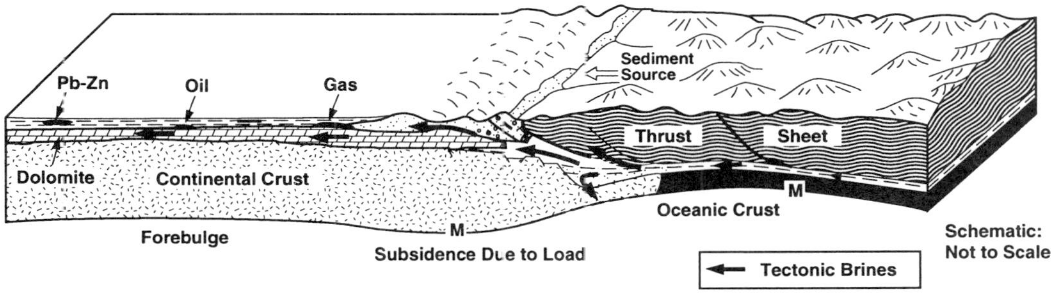

Modified after Oliver (1986)

Figure 76. Block diagram of continental collision. Continental margin sediments are initially porous and filled with seawater and seawater-derived brines. As the thrust sheet progresses, these fluids, which amount to one-third to one-half of the sediment volume, are expelled and move laterally onto the continent (arrows), producing oil and gas accumulations, sedimentary ore deposits, and dolomitizing any limestones which they encounter.

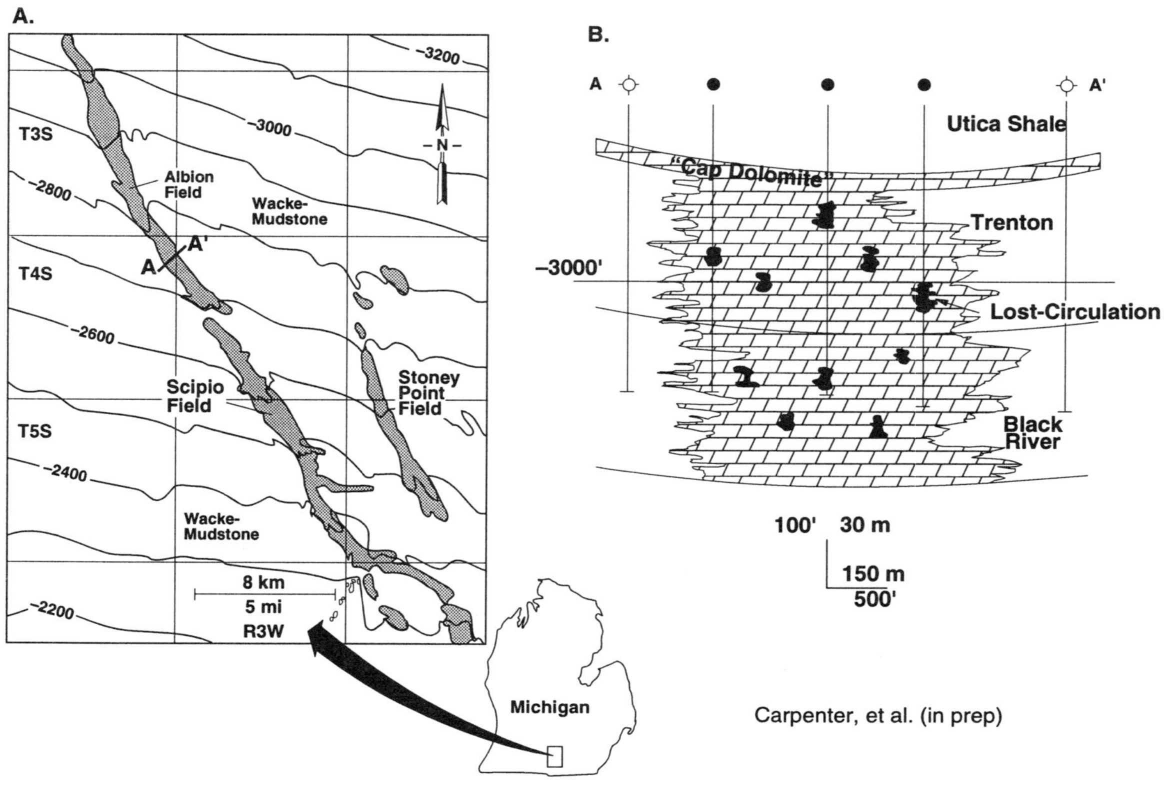

Figure 77. **(A)** Index map showing locations of Albion-Scipio and Stoney Point fields, which are developed along dolomitized strike-slip faults in the Ordovician Trenton Formation, southern Michigan. Fractured and vuggy dolomite, developed along the faults, form long, linear, porous reservoirs which are bounded on either side by tight limestone. Structure contour map on top of the Trenton shows a uniform regional dip and a small synclinal depression over the long, linear Albion-Scipio trend. Synclines above fields were caused by dissolution associated with dolomitization followed by downward collapse of overlying formations. **(B)** Cross section A-A' shows abrupt, almost vertical limestone-dolomite transition. Note syncline on top of field, caused by dissolution associated with dolomitization and downward collapse of overlying formations. Lost circulation zones (shown in black) occur at many depths in the field.

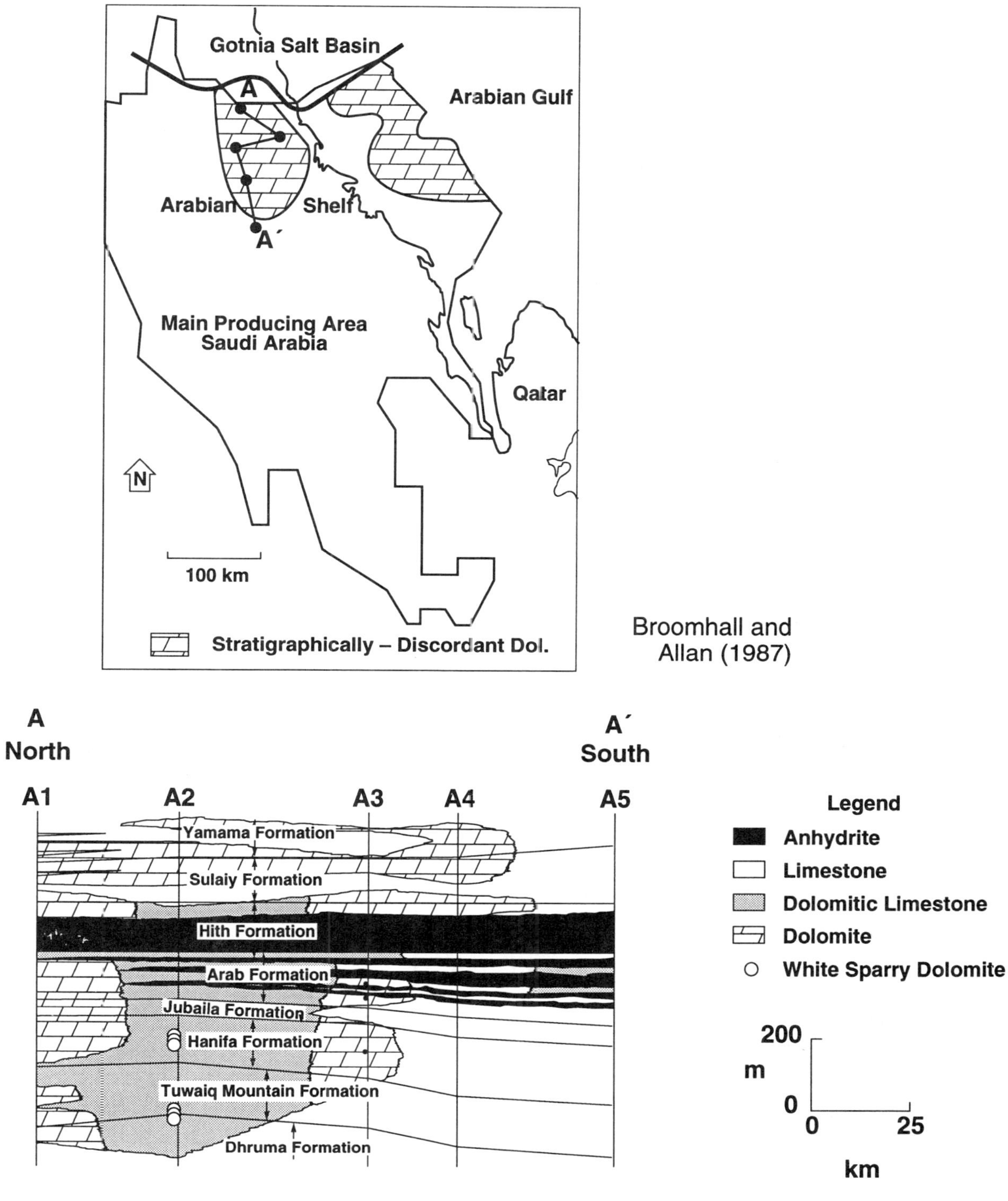

Figure 78. Basemap and cross section showing distribution of thick, tabular dolomite bodies, distributed over hundreds of square miles, which formed by lateral flow of evaporitic brines out of the Gotnia Salt Basin into Jurassic and Early Cretaceous shelf carbonates, eastern Saudi Arabia.

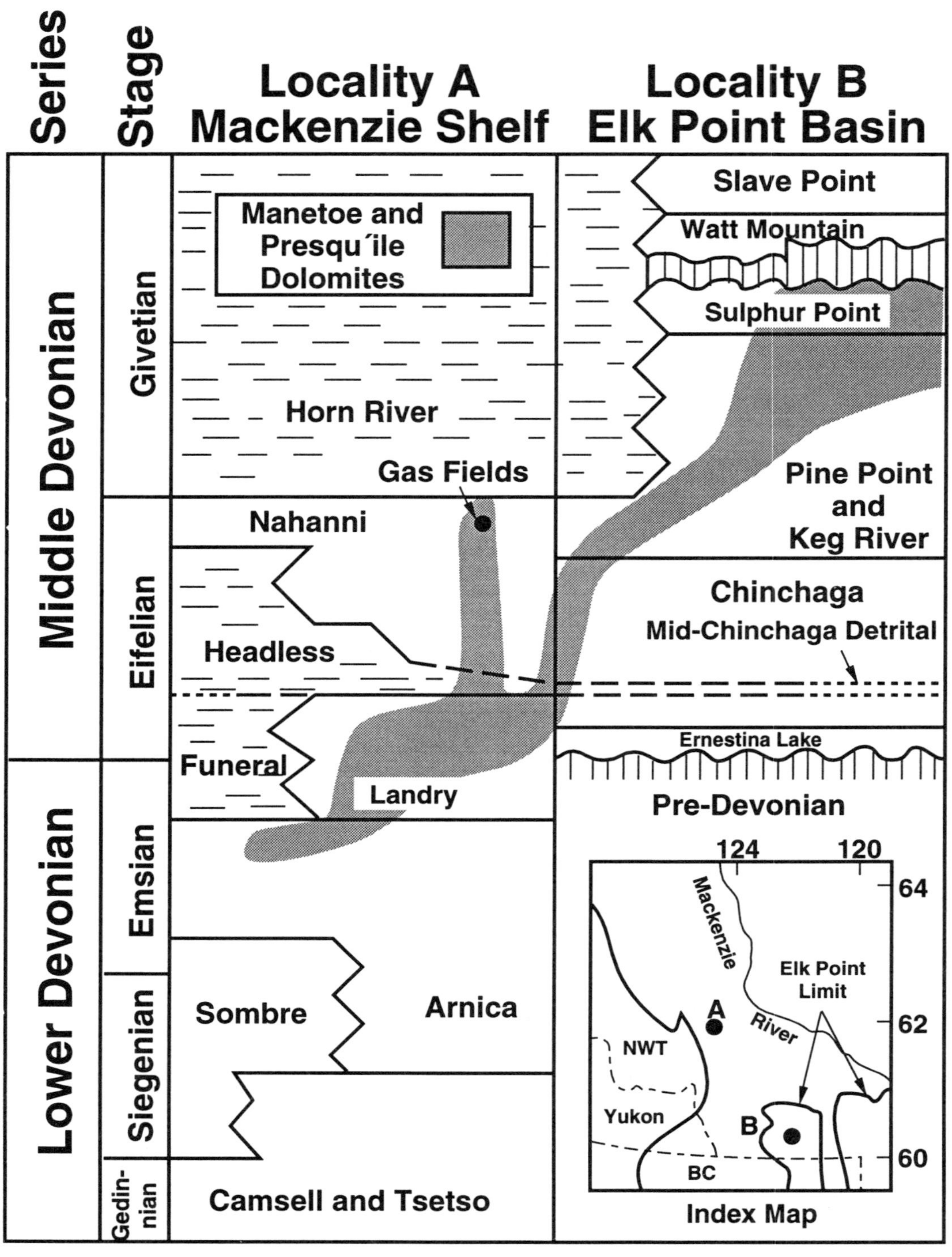

Morrow et al. (1986)

Figure 79. Stratigraphic column which illustrates how late, coarsely crystalline dolomites of the Manetoe and Presqu'ile facies (shaded area) climb section in a southeasterly direction from the Lower to Middle Devonian within carbonates of the Mackenzie shelf, North West Territories, and within carbonates of the Elk Point Basin, North West Territories and northeastern British Columbia.

Burial dolomitization can be advantageous or detrimental to reservoir development. Dolomitization of muddy host rocks commonly produces a tight, interlocking mosaic with low porosity and permeability, whereas dolomitization of grainstones often preserves or enhances primary porosity. For example, burial-dolomitization of channeled debris flows within the Bone Spring Fm of the Delaware Basin produces excellent porosity, whereas dolomitization of the surrounding basinal mudstones results in the seal (Wiggins and Harris, 1985). Porosity developed as a result of dolomitization is often preserved during further burial because dolomite is less susceptible than calcite to porosity loss with depth (Schmoker and Halley, 1982). Partial dolomitization followed by dissolution of undolomitized limestone and bioclasts commonly leads to development of moldic and vuggy porosity (Randazzo and Zachos, 1983). Examples of partial limestone or bioclast dissolution, which lead to important economic reservoirs, include the Devonian of the Alberta Basin, where stromatoporoids have been removed, and the Permian of West Texas, where fusulinids have been removed. Dissolution of evaporites in the burial environment can also lead to collapse of the dolomite reservoir and porosity enhancement. Traps can be either structural or stratigraphic, and sealed by shales, evaporites, or muddy carbonates.

RECOGNITION

Because burial dolomitization normally occurs after deposition and lithification, dolomite typically cross-cuts depositional facies and, in many occurrences, formation boundaries as well. Completely dolomitized rocks are commonly medium to very coarsely crystalline (crystal sizes range from 150 µm to >> 1 mm). They are commonly composed of anhedral to subhedral gray or brown dolomite (recrystallized host rock) which is overlain by or grades into vug- and fracture-lining white dolomite with euhedral terminations (direct precipitates from solution). The white dolomite commonly displays a baroque or saddle texture (curved crystal faces and undulatory extinction). Its white opaque color is caused by an abundance of two-phase aqueous fluid inclusions. Coarse euhedral rhombs of burial dolomite are also found scattered through limestone host rocks, but this type of dolomite does not contribute to reservoir development.

Burial dolomite generally has more **negative $\delta^{18}O$** compositions than earlier-formed dolomite, as a result of its higher temperature of precipitation. It commonly contains **two-phase aqueous fluid inclusions with high salinities and homogenization temperatures >50°C** and may also contain **petroleum inclusions**, readily identifiable by their fluorescence properties, which indicate dolomite precipitation occurred during active petroleum migration. The **Sr isotopic compositions of many (but not all) burial dolomites are radiogenic**, indicating that the dolomitizing brines were allochthonous and interacted with K-feldspars in siliciclastic sediments or basement rocks before entering the carbonate units which they dolomitized. Many burial dolomites are **enriched in Fe and Mn** for the same reason, and **luminesce dull orange** under cathodoluminescent illumination. Burial dolomite is often associated with accessory minerals common to Mississippi Valley-type mineralization, e.g., sphalerite, galena, fluorite, barite, and marcasite.

TRENTON FORMATION

The Albion-Scipio Field in the Ordovician Trenton Formation in southern Michigan is a good example of a burial dolomite reservoir produced by cross-formational flow of residual evaporite brine through a near-vertical fault and fracture system (Carpenter, et al., in prep.). The Trenton Formation in the area is composed of tight wacke-mudstones that were deposited on a regional, northward dipping ramp (Fig. 77A). The Trenton is overlain by the Utica Shale, reef and interreef carbonates of the Niagara Formation, and Salina Formation evaporites (Fig. 80). Repeated reactivation of a Precambrian left-lateral wrench fault system occurred throughout the early and middle Paleozoic, opening a conduit through which Salina residual evaporite brines were able to flow downward, driven by their high density. This resulted in a long, linear, porous dolomite reservoir in the Trenton Formation, <1 mi wide by 30 mi long, which is bounded on either side by tight limestone. Dissolution of host limestone associated with dolomitization caused downward collapse of overlying formations, and formed a syncline with up to 50 ft of relief on the top of the field (Fig. 77B). The collapsed interval can be traced upward as far as the Devonian section, where it dies out into flat-lying beds. Fractures in the Utica Shale, which now acts as the top seal to the reservoir, must have healed when fault movement ceased. Tight, undolomitized wacke-mudstones act as the lateral seal. This results in a very subtle reservoir which contains very significant reserves (>120 MMBO recoverable).

Reservoir dolomite is composed of coarsely crystalline dolomitized host rock that is fractured, vuggy, and cavernous. Fractures and vugs contain white saddle dolomite. Host (replacive) and saddle dolomites have similar ^{18}O-depleted compositions (Fig. 81A) and contain two-phase aqueous fluid inclusions with homogenization temperatures (Th) > 100°C and final melting temperatures (Tm) as low as -42°C (Fig. 81B). Isotopic and fluid inclusion analyses indicate dolomite precipitation from a very saline Na-Ca-Mg-Cl brine during burial. Chemical analyses of present-day Trenton formation waters produced from the Albion-Scipio Field show that they are composed of residual evaporite brine which has precipitated halite, mixed with lower salinity water - probably Ordovician connate water plus later fluids from other sources (Fig. 82). Thus, both the extremely high fluid inclusion salinities and the formation water geochemistry point to Salina brine as a major component of Trenton Formation water. Both host and saddle dolomites contain two-phase hydrocarbon inclusions, which point to late dolomitization during hydrocarbon migration.

Strontium isotopic compositions of Trenton dolomites fall between published values for Trenton-aged Ordovician seawater and Salina-aged Silurian seawater (Fig. 83). This provides further support for dolomitization during mixing of Salina brine with original Trenton interstitial water. The non-radiogenic Sr isotopic compositions of the dolomite argue against invasion of dolomitizing fluids from underlying Cambro-Ordovician siliciclastic aquifers or Precambrian crystalline basement.

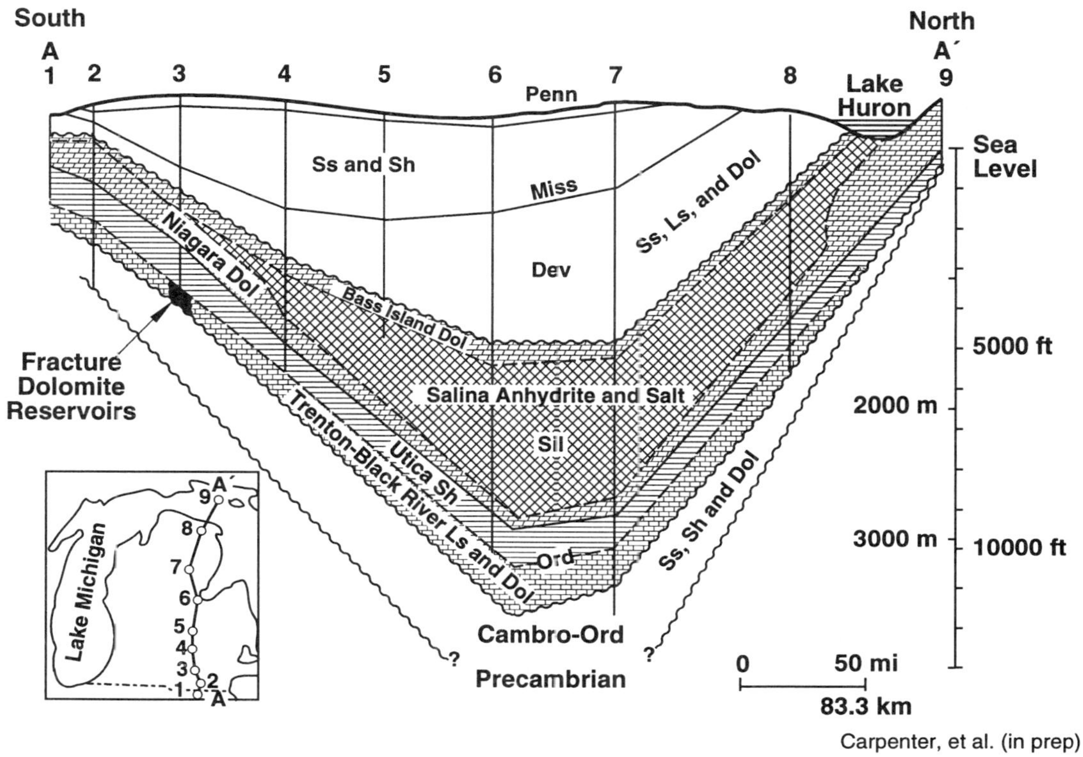

Figure 80. North-South cross section showing stratigraphic relationships in the Michigan Basin. Note especially the position of fracture-dolomite reservoirs at Albion-Scipio and Stony Point fields (Location 3) and the development of Silurian evaporites upsection.

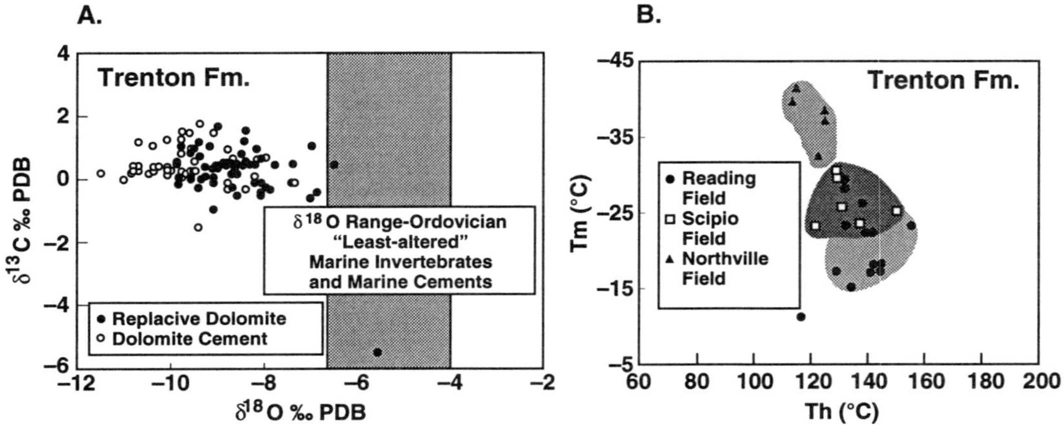

Carpenter, et al. (in prep)

Figure 81. **(A)** Replacive dolomite and dolomite cement in the Trenton Formation have similar $\delta^{18}O$ vs $\delta^{13}C$ values, suggesting dolomitization of both was accomplished by same fluid. ^{18}O-depleted compositions indicate high temperature origin. **(B)** Th and Tm data for primary fluid inclusions trapped along growth bands in dolomite cement from several fracture dolomite fields in Southern Michigan indicate precipitation at temperatures > 100°C from a very saline Na-Ca-Mg-Cl brine. Tightly clustered data within each sample indicate stretching and hydrofracturing of inclusions did not occur.

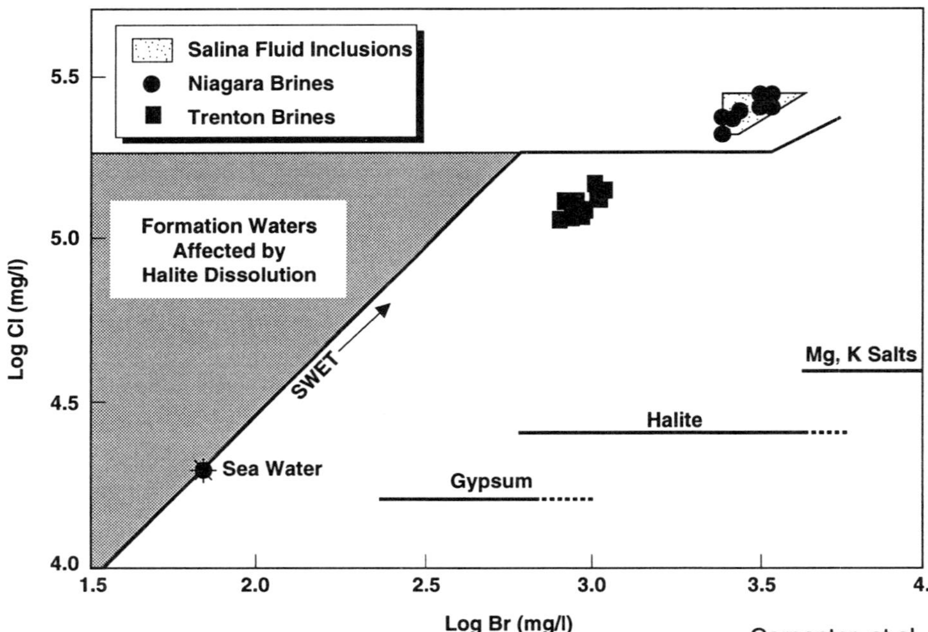

Carpenter, et al. (in prep)

Figure 82. Plot of chloride vs bromide contents of present-day waters collected from the Trenton and Niagara formations in the vicinity of the Albion-Scipio Field. Niagara Formation waters are residual evaporite brines that were expelled from Salina evaporites that envelop Niagaran pinnacle reef reservoirs. They have the same Cl and Br contents as fluid inclusions in Salina halite. Trenton Formation waters plot to the right of the point of initial halite precipitation on the Seawater Evaporation Trend (SWET), indicating that they contain a residual evaporite brine, but below the SWET, indicating that they have mixed with water of lower salinity. Since the only significant nearby halite deposits occur in the Salina, Silurian brines must have percolated downward into the Trenton fracture reservoir in Albion-Scipio Field. Also shown on the figure are Br concentration ranges associated with precipitation of gypsum, halite, and magnesium and potash salts.

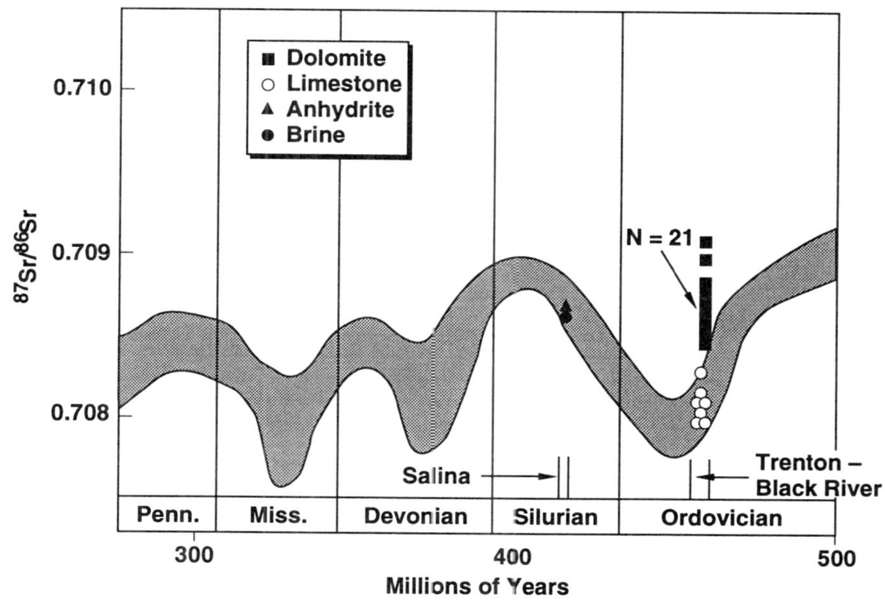

Carpenter et al. (in prep.)

Figure 83. Strontium isotopic compositions of Trenton limestone and dolomite and Salina anhydrite and brine. Trenton limestone has Trenton-age Ordovician seawater values and Salina anhydrite and brine have Salina-age Silurian seawater values. 21 of 23 Trenton dolomite samples have values closest to Salina-age seawater, supporting dolomitization by Salina brine.

SAUDI ARABIAN CRETACEOUS AND JURASSIC FORMATIONS

Upper Jurassic and Lower Cretaceous Formations in Saudi Arabia provide an excellent example of burial dolomitization by lateral flow of residual evaporite brines out of a salt basin during the normal course of burial and basin compaction (Broomhall and Allan, 1987). The Tuwaiq Mountain, Hanifa, Jubaila, Arab, Hith, and Sulaiy Formations in Eastern Saudi Arabia (Fig. 84) form reservoirs composed of pelletal, oolitic, and bioclastic shoal lime packstones and grainstones sealed by tight lime mudstones or supratidal anhydrites. Around the rim of the Gotnia Salt Basin (Fig. 85), regional, stratigraphically discordant dolomitization has enhanced porosity and permeability in the carbonates, has locally led to solution-collapse brecciation of the carbonate reservoirs, and has in many places breached the anhydrite and lime mud seals. Maps and cross-sections constructed from log and core data from 50 wells display plumes of dolomitization which are best developed in two salients adjacent to the Gotnia Basin and decrease in thickness southward, until they disappear entirely 75 to 100 mi from the basin margin (Figs. 86 and 87). This distribution immediately suggests dolomitization by compaction-driven brines which migrated out of the basin. Although the dolomite bodies are many hundreds of feet thick and are distributed over thousands of square miles, few structures in the dolomitized salients produce commercial quantities of oil and gas. Leaching associated with dolomitization also created solution-collapse breccias which breach lime mud and evaporite seals in many localities (see, for example, Fig.87B, locality B-2). While abundant evidence of petroleum, in the form of "dead oil" and petroleum fluid inclusions, is observed within dolomitized

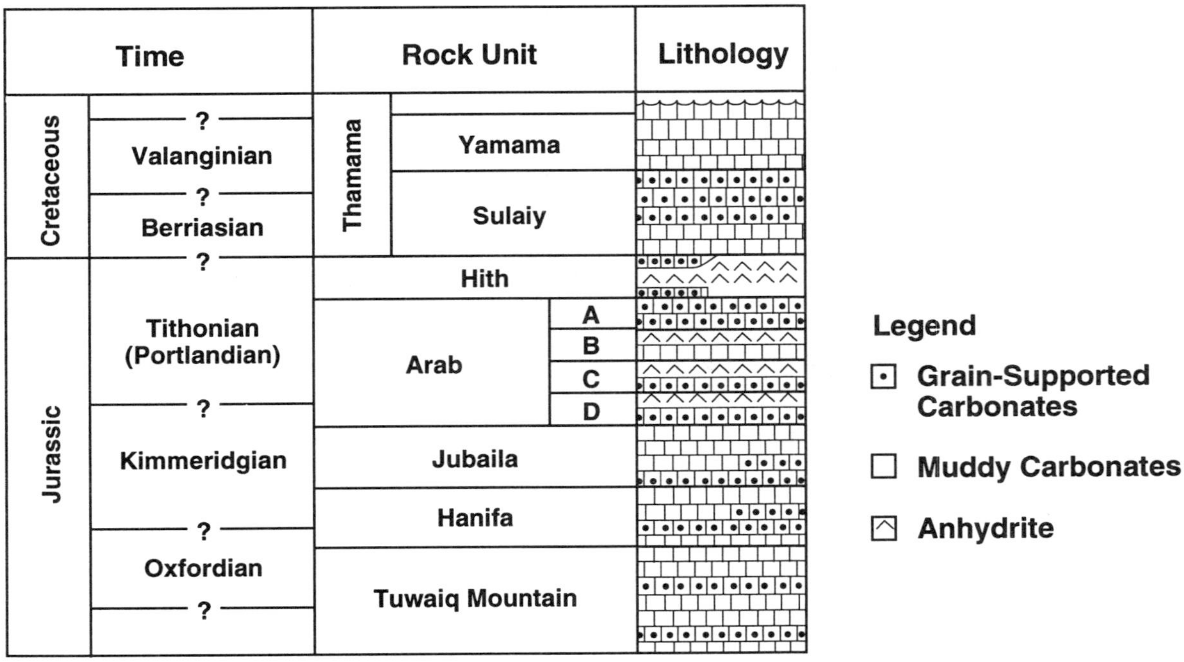

Broomhall and Allan (1987)

Figure 84. Age, stratigraphic relationships, and lithologies of Middle Jurassic to Early Cretaceous carbonate formations, eastern Saudia Arabia.

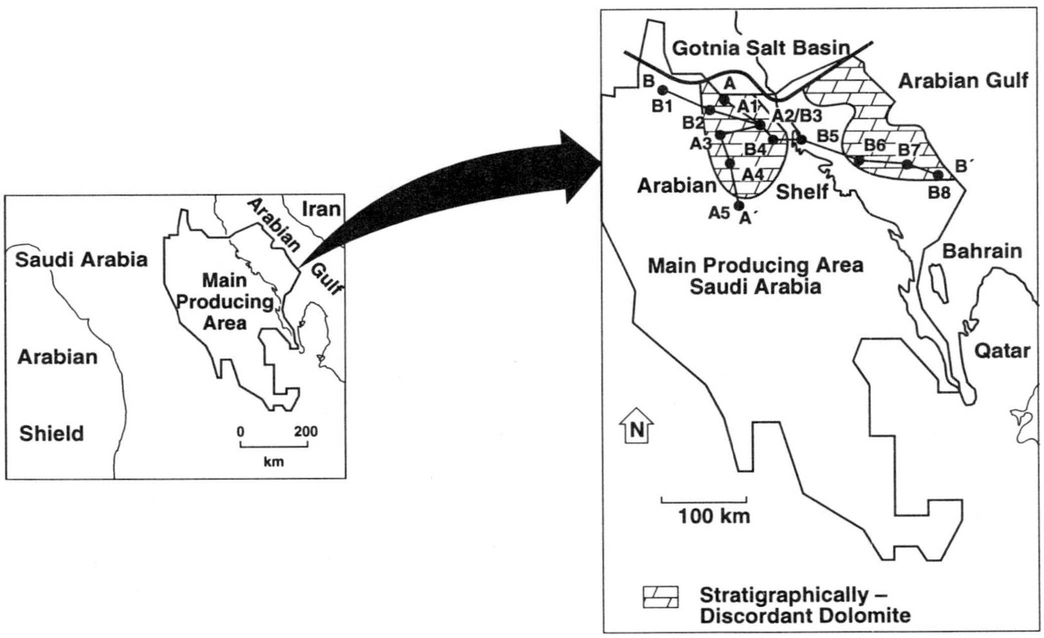

Modified after Broomhall and Allan (1987)

Figure 85. Index map of main producing area in eastern Saudi Arabia, showing general distribution of two salients of stratigraphically discordant dolomite and approximate positions of Mesozoic Arabian Shelf and Gotnia Salt Basin.

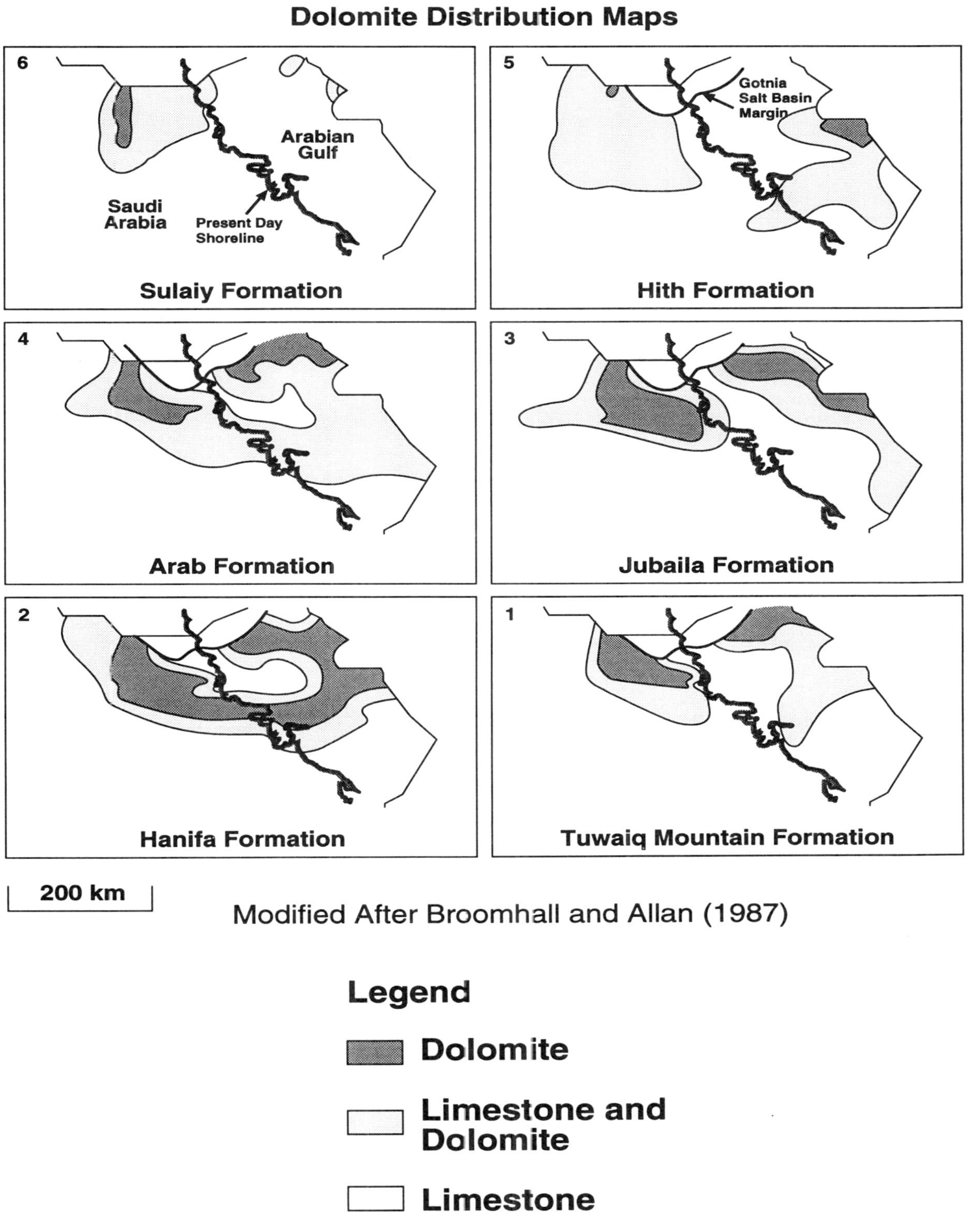

Figure 86. Dolomite distribution maps for Tuwaiq Mountain through Sulaiy formations. Boundary lines delineate northern and eastern borders of main producing area (see Fig.85). Nos. 1 through 6 depict progressively younger time slices.

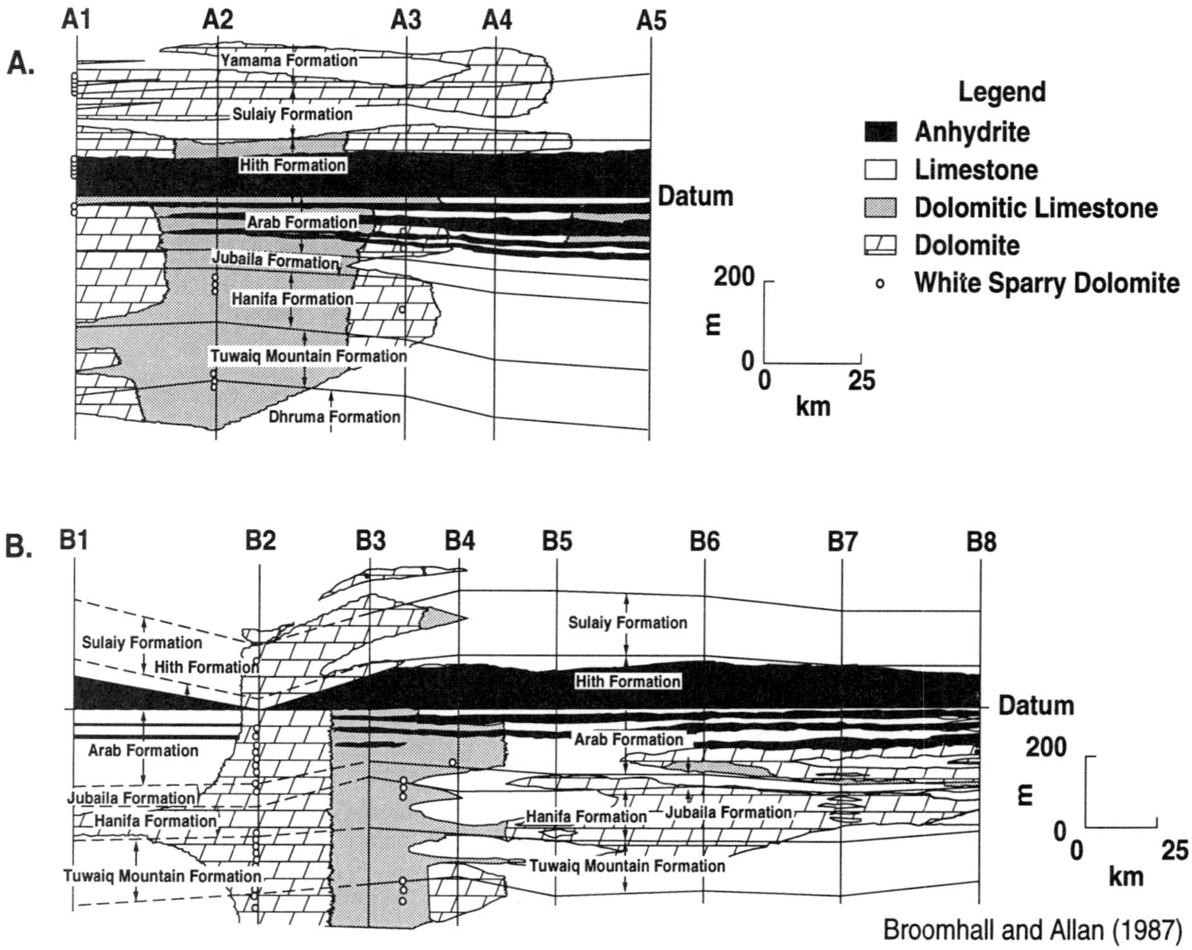

Figure 87. Stratigraphic cross sections A-A' and B-B' from Fig. 85, showing stratigraphic distribution of dolomite. Best dolomite development is adjacent to Gotnia Basin. Dolomite decreases in thickness away from basin, suggesting dolomitization by brines which migrated out of basin. Note extreme vertical exaggeration and large areal extent of dolomite units.

areas, the solution collapse breccias have allowed petroleum to escape upward into overlying formations and, in some areas, all the way to the surface.

Analyses of formation waters collected from within the dolomitized salients indicate they are composed largely of brine that was evaporated past the halite-precipitation point (Fig. 88A). A plot of Ca/Mg activity ratio vs. temperature for the same formation waters, however, indicates they have lost much of their Mg and are now in equilibrium with both calcite and dolomite (Fig. 88B). The lost Mg was apparently consumed by the dolomitization of limestone along the brine migration pathway. Final melting temperatures of fluid inclusions in the dolomite reveal that entrapped fluids are highly concentrated brines with salinities >6 times that of seawater (Fig. 89). Fluid inclusion homogenization temperatures reveal that dolomitization occurred at >100°C. $\delta^{18}O$ compositions also reflect temperature of dolomitization and vary sympathetically with homogenization temperature. More positive $\delta^{18}O$ compositions are associated with lower Th values and more negative $\delta^{18}O$ compositions are associated with higher Th values (Fig. 90A and 90B). While dolomitization in both the eastern and western salients occurred during burial, the generally lower temperatures and more positive $\delta^{18}O$ compositions in the western salient suggest dolomitization occurred somewhat earlier there at shallower burial depths. Replacive and saddle dolomites have similar $\delta^{18}O$ compositions, suggesting both formed from the same fluid (Fig. 90B). Many dolomite samples contain petroleum fluid inclusions as well, indicating that dolomitization occurred contemporaneously with petroleum migration.

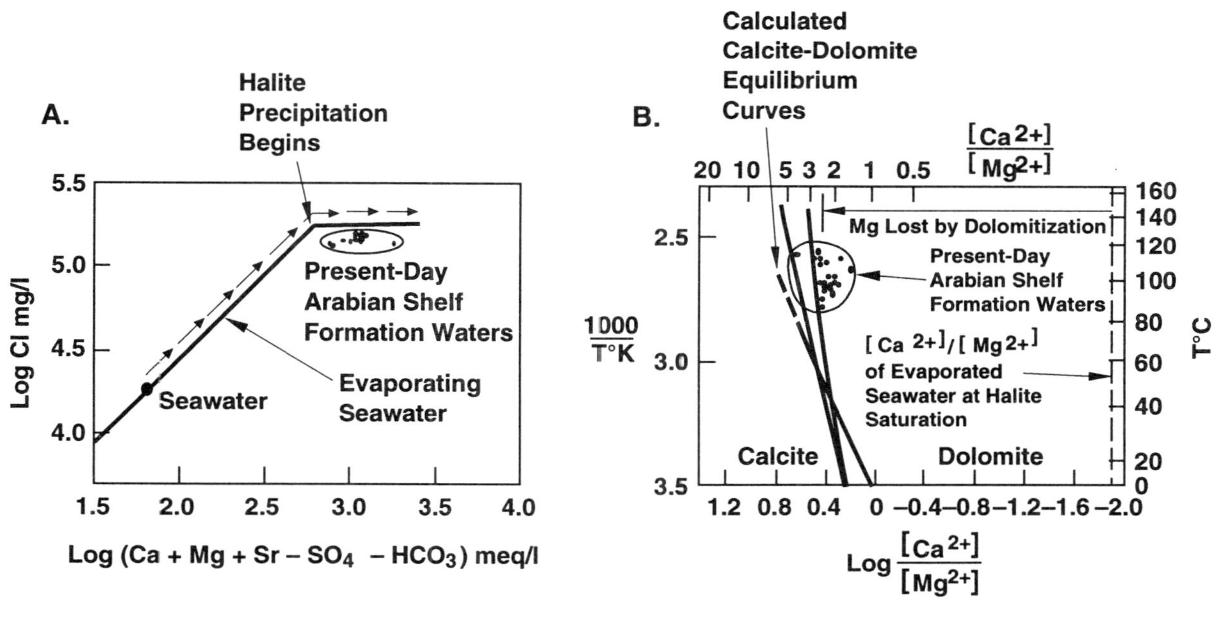

Broomhall and Allan (1987)

Figure 88. **(A)** Log Ca+Mg+Sr-SO_4-HCO_3 vs log Cl plot indicates modern Arabian Shelf formation waters are composed largely of brine evaporated past the halite-precipitation point. **(B)** Ca/Mg vs temperature plot for same formation waters shown in Fig. 88A indicates they have lost much of their Mg, presumably by dolomitizing limestones, and are now in equilibrium with both calcite and dolomite.

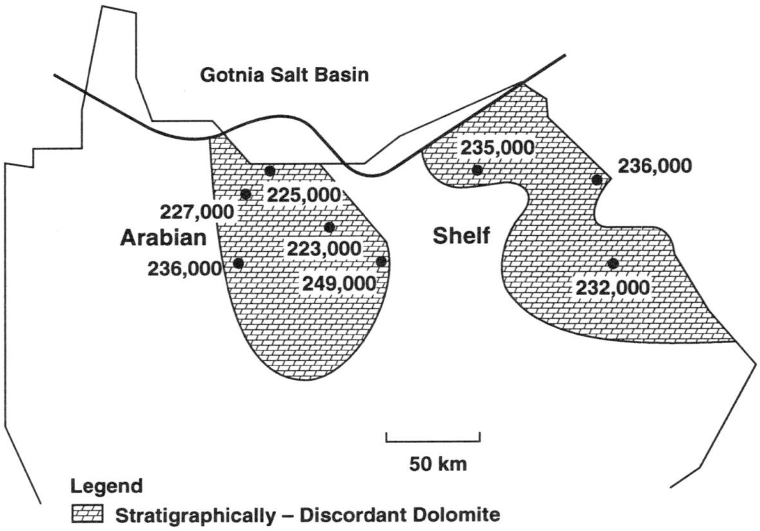

Broomhall and Allan (1987)

Figure 89. Fluid inclusion salinities in saddle dolomite cement (ppm NaCl equivalents). Each number represents the mean value for an oil field.

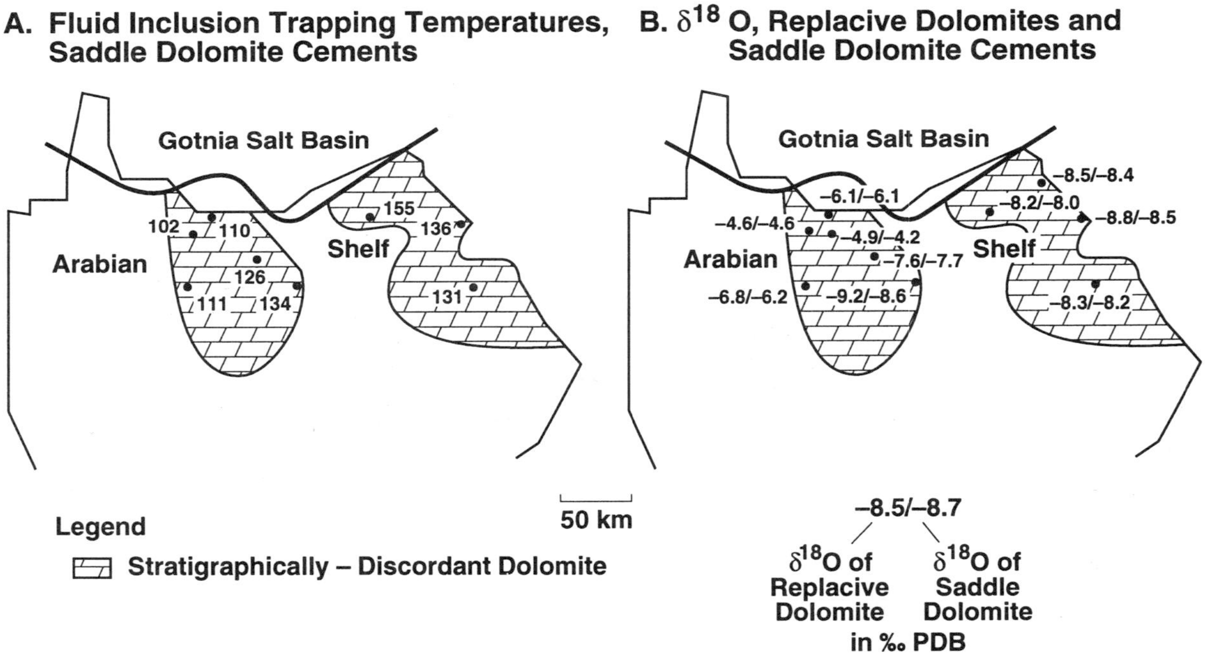

Broomhall and Allan (1987)

Figure 90. (**A**) Fluid inclusion trapping temperatures in saddle dolomite cement. Each number represents the mean value for an oil field. (**B**) Regional variation in oxygen isotopic composition of replacive dolomite and saddle dolomite cement. $\delta^{18}O$ compositions reflect temperature of dolomitization and vary sympathetically with fluid inclusion trapping temperatures. Each number represents the mean value for an oil field. Both replacive dolomite and saddle dolomite cement have similar values within each field, implying a common origin.

NISKU FORMATION

The Late Devonian (Frasnian) Nisku Formation in the Alberta Basin, Canada, provides an example of pervasive dolomitization during burial in which the fluid sources and hydrology responsible for dolomitization are not yet well-understood. This case study is presented as an example of a situation which is commonly encountered in the "real world": geochemistry has provided constraints on the timing and mechanism of dolomitization, but has not provided unequivocal answers.

Nisku sediments were deposited as a carbonate shelf sequence surrounding the Winterburn shale basin during a major regressive cycle (Fig. 91). Pinnacle reefs developed in a deeper water ramp environment downslope from the shelf edge and were eventually encased in peritidal carbonates and subtidal shales (Fig. 92). The reefs are coral mudmounds, up to 250 ft thick and 1 1/4 mi in diameter (Machel and Anderson, 1989). Although depositional dip was toward the northwest, dolomitization increases toward the southwest, along present-day structural dip (Fig. 93). Many reefs show preferential dolomitization in their southwestern quadrants, further suggesting that dolomitizing fluids migrated from down structural dip (Machel, 1986). Significant porosity and permeability is limited to reefs. The best reservoir characteristics are found in completely dolomitized reefs, which produce from moldic, vuggy, intercrystalline, and fracture porosity in fine to medium crystalline dolomite (Anderson and Machel, 1989). The lateral seal on reservoirs is tight off-reef argillaceous carbonate, whereas the top seal is argillaceous siltstone and intertidal to supratidal carbonate of the overlying Calmar and Blue Ridge formations. The oil is divided between many individual reef pools, with reserves for the trend estimated at 1.05 BBO recoverable (Anderson and Machel, 1989).

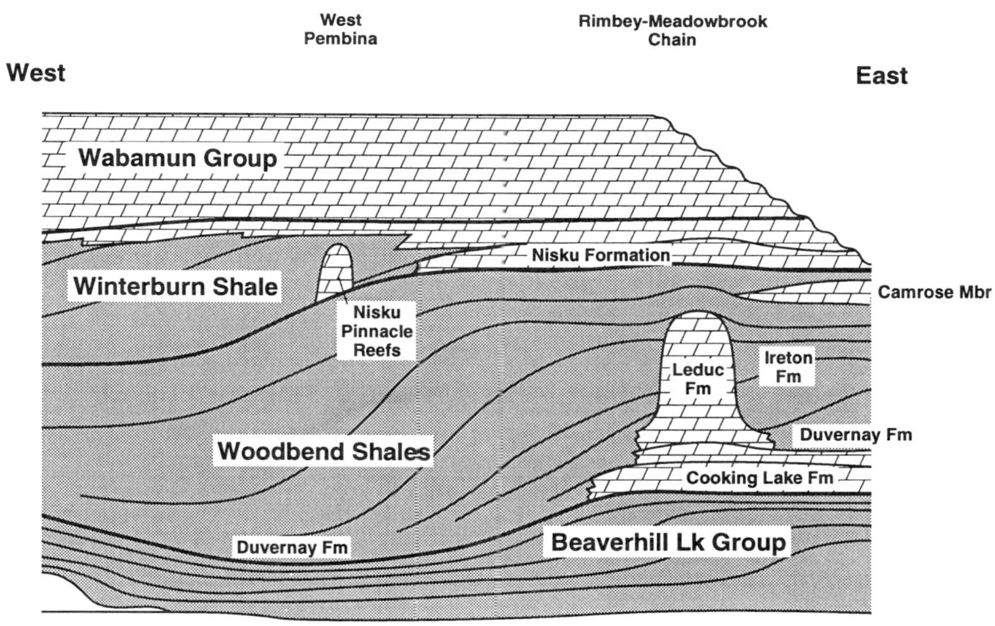

Modified after Stoakes and Wendte (1987)

Figure 91. Cross section showing stratigraphic relationship of Nisku pinnacle reefs to other formations in Upper Devonian section, central Alberta. The Duvernay, a black-colored carbonate unit, is regarded as a regional source rock.

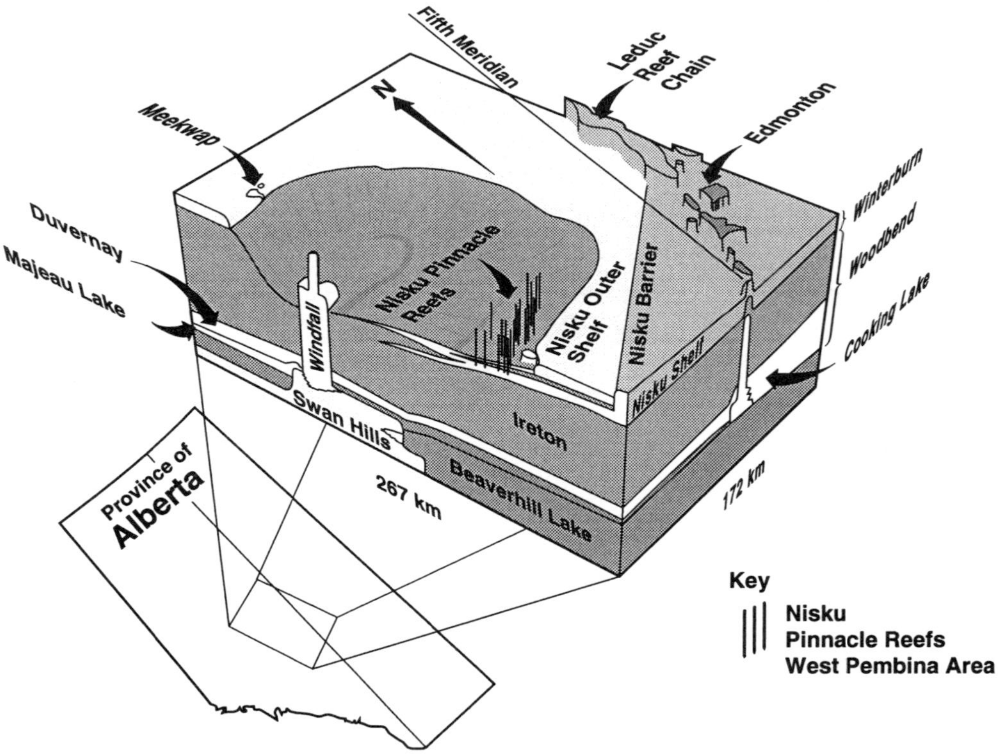

Exploration Staff, Chevron Std. Ltd. (1979)

Figure 92. Block diagram projection, Upper Winterburn Basin, as viewed from the southwest. Pinnacle reefs developed in a deeper water ramp environment downslope from Nisku shelf edge and were later encased in peritidal carbonates and basinal shales.

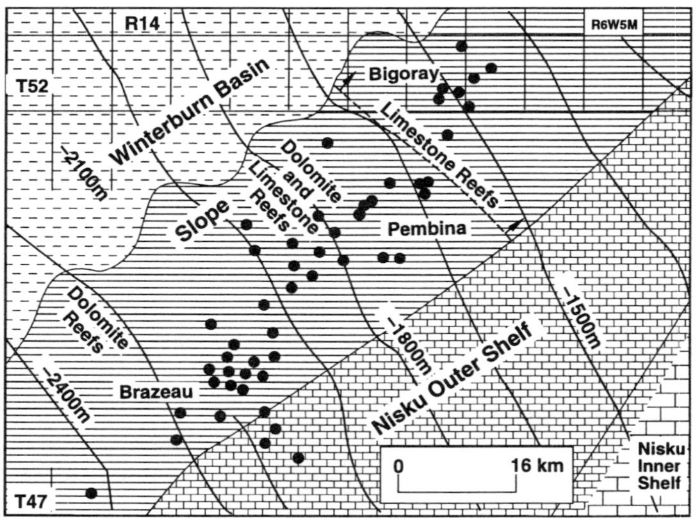

Modified after Moore (1989) and Anderson and Machel (1989)

Figure 93. Map of Nisku pinnacle reef trend, showing positions of inner shelf, outer shelf, slope and basin. Although depositional dip was toward the northwest, dolomitization increases toward the southwest, along present-day structural dip. This suggests that dolomitizing fluids were *not* expelled from adjacent basinal shales, but rather migrated up structural dip during burial. This contention is further supported by a tendency for structurally downdip flanks of individual pinnacles to be selectively dolomitized.

Several generations of dolomite have been identified, including an early generation of finely crystalline dolomite, an intermediate generation of pervasive matrix dolomite and late generations of white saddle dolomite and limpid dolomite that line or fill molds and vugs. The pervasive dolomite is by far the most important to reservoir development and displays sucrosic (euhedral) to xenotopic (interlocking anhedral mosaic) textures. Machel and Anderson (1989) assumed dolomitization by marine connate fluids of Devonian age, which had $\delta^{18}O$ compositions 3 to 4‰ lighter than modern seawater, (Popp, et al., 1986a and b), and calculated a temperature of 40-50°C for pervasive matrix dolomitization. Because this calculation is based on an assumed water value, it is speculative in nature. Several other water compositions (e. g., isotopically heavier brines or isotopically lighter mixed waters) are equally plausible. When one plots Anderson and Machel's data on the types of isotope crossplots presented as Figs. 19 and 21 in Part I, it becomes apparent that carbon and oxygen isotope values are non-definitive. The range of reported $\delta^{18}O$ values for Nisku pervasive dolomite, -2.2 to -6.2‰ PDB, fall mostly within the region of overlap between low temperature and high temperature dolomites (Fig. 94A) and straddle the $\delta^{18}O$ range for Devonian brachiopods and marine carbonate cements (Fig. 94B). This indicates that pervasive dolomite could have formed either at the surface or during shallow burial.

Both the pervasive matrix dolomite and the white saddle dolomite contain radiogenic Sr and fall above the Sr isotope seawater curve (Fig. 95), indicating dolomitizing fluids interacted with siliciclastic sediments or basement prior to dolomitization. This supports an allochthonous source for the fluids which is consistant with a burial origin. Upward increase in Sr and decrease in Mn in reef dolomite reflects changes in trace element partitioning during dolomitization and fluid evolution. Machel and Anderson (1989) argue that these Sr and Mn trends indicate dolomitizing fluids entered Nisku reefs from below and migrated upward - a flow direction which, if correct, is most consistent with burial dolomitization. Fluid inclusion data are only available for white saddle dolomite. Homogenization temperatures range from 112 to 175°C and salinities range from 8.5 to 18 wt% NaCl equivalents, indicating precipitation in the deep subsurface from concentrated brines (Anderson, 1985). However, since white saddle dolomite appears to be a late-formed phase, it may have little relationship to the pervasive matrix dolomite which forms the reservoirs.

The burial history of the Nisku Formation is shown in Fig. 96. Burial to 1000 m occurred in the Mississippian. In the Pennsylvanian through early Jurassic, southwestward tilting occurred and much of the post-Nisku section was eroded in the East. Tilting and erosion were accompanied by subvertical faulting that affected the entire Paleozoic section. During uplift of the Rocky Mountains, in the Cretaceous and Early Tertiary, the Nisku was buried under a wedge of sediments that thickened toward the southwest.

At least three dolomitization mechanisms consistent with the geochemical data could have operated within this framework. (1) Machel (1985, 1986) argued that pervasive matrix dolomitization was accomplished by connate marine waters, expelled during burial from compacting basinal shales, which migrated updip and dolomitized Nisku reefs from the bottom up. (2) Anderson (1985) argued that Machel's model has a severe mass balance shortfall: insufficient Mg is available from shale dewatering to explain the amount of dolomite observed. Anderson (1985) speculated that a gravity-driven hydrologic system may have formed during Late Paleozoic to Jurassic cratonic tilting and led to either mixed-water dolomitization of the Nisku reefs, or development of a Kohout convective half-

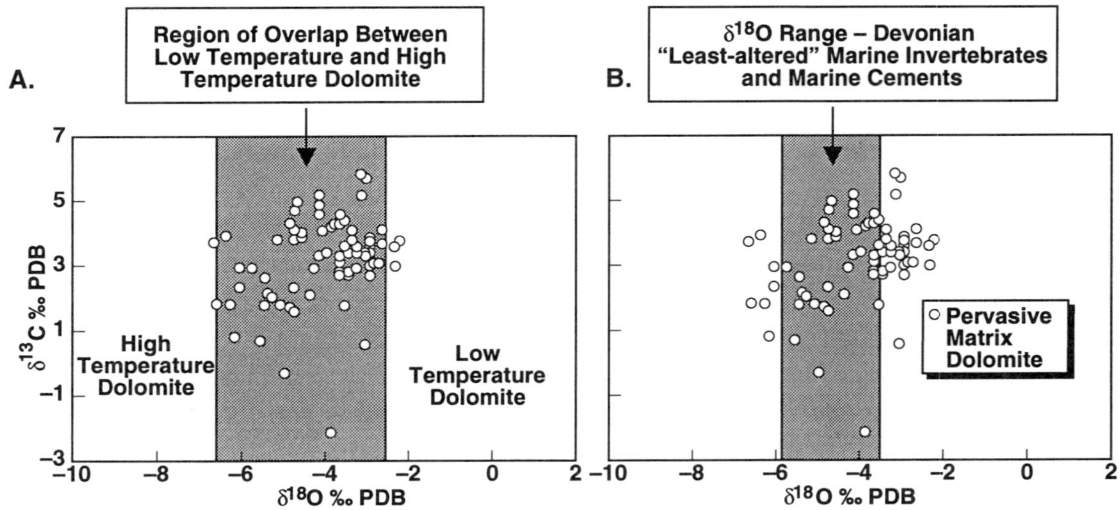

Data from Anderson (1985) and Machel (1985)

Figure 94. **(A)** $\delta^{18}O$ vs $\delta^{13}C$ cross plot for Nisku pervasive dolomite compared to fields for low temperature and high temperature dolomites. Range of reported $\delta^{18}O$ values falls mostly within region of overlap between low temperature and high temperature dolomite, indicating pervasive dolomite could have formed either at the surface or during shallow burial. **(B)** $\delta^{18}O$ vs $\delta^{13}C$ cross plot for Nisku pervasive dolomite compared to Devonian brachiopods and marine carbonate cements. Most pervasive dolomite has $\delta^{18}O$ values similar to or isotopically heavier than Devonian marine carbonates, suggesting an early, nearsurface origin.

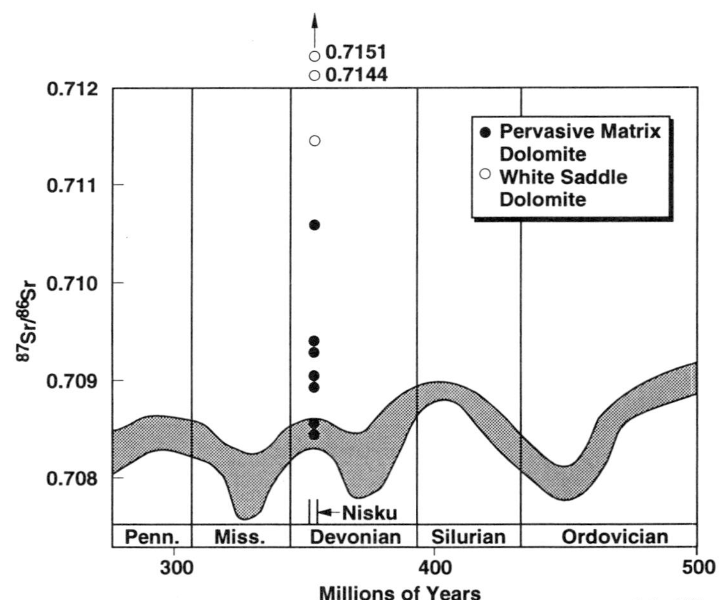

Modified after Anderson (1985)

Figure 95. $^{87}Sr/^{86}Sr$ ratios for Nisku dolomite compared to Sr isotope age curve. Most pervasive dolomite and all white saddle dolomite samples are enriched in ^{87}Sr, indicating dolomitizing fluids interacted with siliciclastic sediments or basement prior to dolomitization and therefore are allochthonous. This is consistent with burial origin for dolomite.

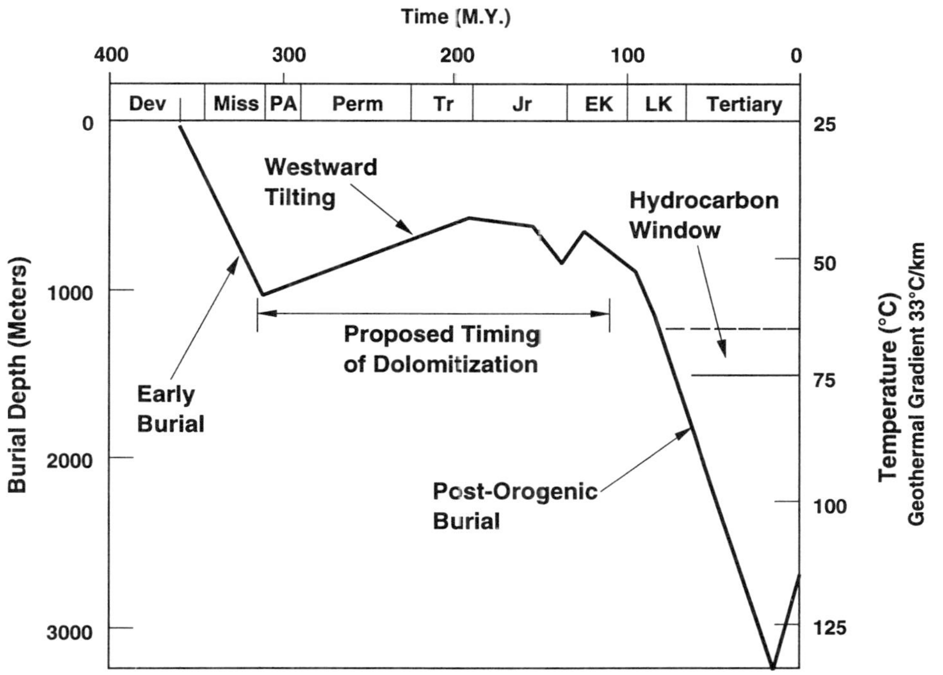

Modified after Anderson (1985)

Figure 96. Burial history curve for Nisku Formation, showing estimated timing of dolomitization.

cell that induced thermal circulation of warm seawater through the reefs and dolomitized them. (3) Finally, Machel and Anderson (1989), in a more recent paper, suggested that thermal convection could have circulated underlying Middle Devonian evaporite brines and overlying marine-derived formation waters through the Nisku section along sub-vertical faults to produce the pervasive dolomite. Both Machel and Anderson believe that saddle dolomite, because of its high homogenization temperatures and radiogenic Sr isotopic compositions, must have formed later during deep burial associated with uplift of the Rocky Mountains in the Cretaceous and Early Tertiary.

Dolomite geometry and geochemistry point to a burial dolomitization mechanism for the Nisku reefs. However, the exact mechanism which produced the dolomite and the hydrology controlling its distribution have yet to be identified because of the complex burial, uplift, and hydrologic history of the Alberta Basin.

REFERENCES

REFERENCES

Adams, J.E. and Rhodes, M. L. (1960) Dolomitization by seepage refluxion: Amer. Assoc. Petroleum Geologists Bull., Vol. 44, p.1912-1920.

Adlis, D. S., Grossman, E. L., Yancey, T. E., and McLerran, R. D. (1988) Isotope stratigraphy and paleodepth changes of Pennsylvanian cyclical sedimentary deposits: Palaios, Vol. 3, p. 487-506.

Aharon, P., Kolodny, Y., and Sass, E. (1976) Recent hot brine dolomitization in the "Solar Lake", Gulf of Elat, isotopic, chemical, and mineralogical study: Jour. Geology, Vol. 85, p. 27-48.

Al-Aasm, I. S. and Veizer, J. (1986) Diagenetic stabilization of aragonite and low-Mg calcite, II. Stable isotopes in rudists: Jour. Sed. Petrology, Vol. 56, p. 763-770.

Allan, J. R. and Matthews, R. K. (1982) Isotope signatures associated with early meteoric diagenesis: Sedimentology, Vol. 29, p. 797-817.

Alsharhan, A. S., and Williams, D. F. (1987) Petrography and stable isotope composition of baroque dolomite from the Shuaiba Formation (Lower Cretaceous), Abu Dhabi, United Arab Emirates: Jour. African Earth Sci., Vol. 6, p. 881-890.

Anderson, J. H. (1985) Depositional facies and carbonate diagenesis of the downslope reefs in the Nisku Formation (Upper Devonian), Central Alberta, Canada: unpubl. Ph.D. dissert., Univ. of Texas, Austin, TX, 393 pp.

Anderson, J. H. and Machel, H. G. (1989) The Upper Devonian Nisku reef trend in Central Alberta: in Geldsetzer, H.H.J., James, N.P., and Tebbutt, eds., Reefs, Canada and adjacent areas, Can. Soc. Petrol. Geol., Mem. 13, p. 391-398.

Aulstead, K. L, Spencer, R. J., and Krouse, H. R. (1988) Fluid inclusion and isotopic evidence on dolomitization, Devonian of western Canada: Geochim. Cosmochim. Acta, Vol. 52, p. 1027-1035.

Badiozamani, K. (1973) The Dorag dolomitization model - application to the middle Ordovician of Wisconsin: Jour. Sed. Petrology, Vol. 43, p.965-984.

Banner, J. L., Hanson, G. N., and Meyers, W. J. (1988) Water-rock interaction history of regionally extensive dolomites of the Burlington-Keokuk Formation (Mississippian): isotopic evidence: in Shukla, V. and Baker, P. A., eds., Sedimentology and Geochemistry of Dolostones, Soc. Econ. Paleontologists and Mineralogists Spec. Publ. 43, p. 97-113.

Barker, J. F. and Fritz, P. (1981) Carbon isotope fractionation during microbial alteration of methane: Nature, Vol. 293, p.289-291.

Bay, T. A. (1983) The Silurian of the northern Michigan Basin: in Harris, P. M., ed., Carbonate Buildups - a Core Workshop, Soc. Econ. Paleontologists and Mineralogists Core Workshop No. 4, p. 53-72.

Beaty, D. W. and Landis, G. P. (1990) Origin of the ore deposits at Gilman, Colorado. Part IV: Stable isotope geochemistry: in Beaty, D. W., Landis, G. P., and Thompson, T. B., eds., Carbonate-hosted Sulfide Deposits of the Central Colorado Mineral Belt, Econ. Geol. Monograph No. 7, p. 228-245.

Behrens, E. W. and Land, L. S. (1972) Subtidal Holocene dolomite, Baffin Bay, Texas: Jour. Sed. Petrology, Vol. 42, p. 155-161.

Bellanca, A. and Neri, R. (1986) Evaporite-carbonate cycles of the Messinian, Sicily: stable isotopes, mineralogy, textural features, and environmental implications: Jour. Sed. Petrology, Vol. 56, p. 614-621.

Bethke, C. M. (1986) Hydrologic constraints on the genesis of the Upper Mississippi Valley district from Illinois Basin brines: Econ. Geol., Vol. 81, p. 233-249.

Boles, J. R. (1978) Active ankerite cementation in the subsurface Eocene of Southewest Texas: Contrib. Mineral. and Petrol., Vol. 68, p. 13-22.

Brand, U. (1981) Mineralogy and chemistry of the Lower Pennsylvanian Kendrick fauna, eastern Kentucky, 2. Stable isotopes: Chem. Geol., Vol. 32, p. 17-28.

Brand, U. (1982) The oxygen and carbon isotope composition of Carboniferous fossil components: seawater effects: Sedimentology, Vol. 29, p. 139-147.

Brand, U. (1986) Paleoenvironmental analysis of Middle Jurassic (Callovian) ammonoids from Poland: trace elements and stable isotopes: Jour. Paleontology, Vol. 60, p. 293-301.

Brand, U. (1987) Depositional analysis of the Breathitt Formation's marine horizons, Kentucky, U.S.A.: Trace elements and stable isotopes: Chem. Geol., Vol. 65, p. 117-136.

Brand, U. (1989a) Aragonite-calcite transformation based on Pennsylvanian molluscs: Bull. Geol. Soc. Amer., Vol. 101, P. 377-390.

Brand, U. (1989b) Biogeochemistry of Late Paleozoic North American brachiopods and secular variation of seawater composition: Biogeochemistry, Vol. 7, p. 159-193.

Brand, U. (1990) Brachiopod stable-isotope stratigraphy of tropical Devonian - Carboniferous epeiric seas: Geol. Soc. Amer., Abs. with Prog., p. A115 (abs.).

Brand, U. and Morrison, J. O. (1987a) Diagenesis and pyritization of crinoid ossicles: Can. Jour. Earth Sci., Vol. 24, p. 2486-2498.

Brand, U. and Morrison, J. O. (1987b) Paleoscene #6. Biogeochemistry of fossil marine invertebrates: Geosci. Canada, Vol. 14, p.85-107.

Brand, U. and Veizer, J. (1981) Chemical diagenesis of a multicomponent carbonate system - 2: Stable isotopes: Jour. Sed. Petrology, Vol. 51, p. 987-997.

Brass, G. W. (1976) The variation of the marine $^{87}Sr/^{86}Sr$ ratio during Phanerozoic time: Interpretation using a flux model: Geochim. Cosmochim. Acta, Vol. 40, p. 721-730.

Broomhall, R. W. and Allan, J. R. (1987) Regional caprock-destroying dolomite on the Middle Jurassic to Early Cretaceous Arabian Shelf: SPE Formation Evaluation, Vol. 2, p. 435-441.

Buchardt, B. (1978) Oxygen isotope palaeotemperatures from the Tertiary period in the North Sea area: Nature, Vol. 275, p. 121-123.

Buchardt, B., and Weiner, S. (1981) Diagenesis of aragonite from Upper Cretaceous ammonites: a geochemical case-study: Sedimentology, Vol. 28, p. 423-438.

Buchbinder, L. G., Margaritz, M. and Goldberg, M. (1984) Stable isotope study of karstic-related dolomitization: Jurassic rocks from the coastal plain, Israel: Jour. Sed. Petrology, Vol. 54, p. 236-256.

Budai, J. M., Lohmann, K. C., and Wilson, J. L. (1987) Dolomitization of the Madison Group, Wyoming and Utah Overthrust Belt: Amer. Assoc. Petroleum Geologists Bull., Vol. 71, p. 909-924.

Buelter, D. P. and Guillemette, R. N. (1988) Geochemistry of epigenetic dolomite associated with lead-zinc mineralization of the Viburnum Trend, southeast Missouri: a reconnaissance study: in Shukla, V. and Baker, P. A., eds., Sedimentology and Geochemistry of Dolostones, Soc. Econ. Paleontologists and Mineralogists Spec. Publ. 43, p. 97-113.

Burke, W. H., Denison, R. E., Hetherington, E. A., Koepnick, R. B., Nelson, H. F., and Otto, J. B. (1982) Variation of seawater $^{87}Sr/^{86}Sr$ throughout Phanerozoic time: Geology, Vol. 10, p. 516-519.

Burns, S. J. and Baker, P. A. (1987) A geochemical study of dolomite in the Monterey Formation, California: Jour. Sed. Petrology, Vol. 57, p. 128-139.

Burruss, R. C. (1981) Hydrocarbon fluid inclusions in studies of sedimentary diagenesis: in Hollister, L. S. and Crawford M. L., eds., Short Course in Fluid Inclusions: Applications to Petrology. Mineralogical Assoc. Canada, p. 138-156.

Burruss, R. C. (1987) Diagenetic paleotemperatures from aqueous fluid inclusions: re-equilibration of inclusions in carbonate cements by burial heating: Mineralogical Mag., V. 51, p. 477-481.

Burruss, R. C. (1991) Practical aspects of fluorescence microscopy of petroleum fluid inclusions: in Barker, C. E. and Kopp, O. C., eds., Luminescence Microscopy and Spectroscopy: Qualitative and Quantitative Applications, Soc. Econ. Paleontologists and Mineralogists, Short Course Notes, No. 25, p. 1-7.

Burtner, R. L. (1987) Origin and evolution of Weber and Tensleep formation waters in the Greater Green River and Uinta-Piceance basins, northern Rocky Mountains area, USA: Chem. Geol., Vol. 65, p. 255-282.

Butler, G. P. (1969) Modern evaporite deposition and geochemistry of coexisting brines, the sabkha, Trucial Coast, Arabian Gulf: Jour. Sed. Petrology, Vol. 39, p. 70-89.

Cander, H. S., Kaufman, J., Daniels, L. D. and Meyers, W. J. (1988) Regional dolomitization of shelf carbonates in the Burlington-Keokuk Formation (Mississippian), Illinois and Missouri: Constraints from cathodoluminescent zonal stratigraphy: in Shukla, V. and Baker, P. A., eds., Sedimentology and Geochemistry of Dolostones. Soc. Econ. Paleontologists and Mineralogists Spec. Pub. No. 43, p. 129-144.

Carballo, J. S., Land, L. S., and Miser, D. E. (1987) Holocene dolomitization of supratridal sediments by active tidal pumping, Sugarloaf Key, Florida: Jour. Sed. Petrology, Vol. 57, p. 153-165.

Carpenter, A. B. (1976) Dorag dolomitization model by K. Badiozamani - a discussion: Jour. Sed. Petrology, Vol. 46, p. 258-261.

Carpenter, A. B. (1978) Origin and chemical evolution of brines in sedimentary basins: Oklahoma Geol. Surv. Circ. 79, p. 60-77.

Carpenter, A. B. (1980) The chemistry of dolomite formation I: the stability of dolomite: in Zenger, D. H., Dunham, J. B., and Ethington, R. L., eds., Concepts and Models of Dolomitization. Soc. Econ. Paleontologists and Mineralogists Spec. Pub. No. 28, p. 111-121.

Carpenter, A. B., Allan, J. R., Granath, V. C., Hurley, N. H., and Cumella, S. P. (in prep.) Fracture-controlled dolomitization, Albion-Scipio Field and adjacent areas: Bull. Geol. Soc. Amer.

Carpenter, S. J., and Lohmann, K. C. (1989) $\delta^{18}O$ and $\delta^{13}C$ variations in Late Devonian marine cements from the Golden Spike and Nevis reefs, Alberta, Canada: Jour. Sed. Petrology, Vol. 59, p. 792-814.

Castle, J. W. (1981) Aspects of fluorescence recorded by core photography: Unpublished Chevron report, 10 p.

Cercone, K. R. and Lohmann, K. C. (1985) Early diagenesis of middle Silurian pinnacle reefs, northern Michigan: in Cercone, K. R. and Budai, J. M., eds., Ordovician and Silurian rocks of the Michigan Basin and its Margins, Michigan Basin Geol. Soc. Spec. Paper No. 4, p.109-130.

Cercone, K. R. and Lohmann, K. C. (1987) Late burial diagenesis of Niagaran (Middle Silurian) pinnacle reefs in Michigan Basin: Amer. Assoc. Petroleum Geologists Bull., Vol. 71, p. 156-166.

Cervato, C. (1990) Hydrothermal dolomitization of Jurassic-Cretaceous limestones in the southern Alps (Italy): relation to tectonics and volcanism: Geology, Vol. 18, p. 458-461.

Choquette, P. W. (1971) Late ferroan dolomite cement, Mississippian carbonates, Illinois Basin, U. S. A.: in Bricker, O. P., ed., Carbonate Cements, Johns Hopkins Press, Vol. 19, Baltimore, p. 339-346.

Choquette, P. W. and Pray, L. C. (1970) Geologic nomenclature and classification of porosity in sedimentary carbonates: Amer. Assoc. Petroleum Geologists Bull., Vol. 54, p. 207-250.

Choquette, P. W. and Steinen, R. P. (1980) Mississippian non-supratidal dolomite, Ste. Genevieve Limestone, Illinois Basin: evidence for mixed-water dolomitization: in Zenger, D. H., Dunham, J. B., and Ethington, R. L., eds., Concepts and Models of Dolomitization, Soc. Econ. Paleontologists and Mineralogists Spec. Publ. 28, p.163-196.

Choquette, P. W. and Steinen, R. P. (1985) Mississippian oolite and non-supratidal dolomite reservoirs in the Ste. Genevieve Formation, North Bridgeport field, Illinois Basin: in Roehl, P. O. and Choquette, P. W., eds., Carbonate Petroleum Reservoirs, Springer-Verlag, New York, p. 209-225.

Clayton, R. N., Skinner, H. C. W., Berner, R. A., and Rubinson, M. (1968) Isotopic compositions of recent South Australian lagoonal carbonates: Geochim. Cosmochim. Acta, Vol. 32, p. 983-988.

Collins, S. E., (1980) Jurassic Cotton Valley and Smackover reservoir trends, East Texas, North Louisiana, and South Arkansas: Amer. Assoc. Petroleum Geologists Bull., Vol. 64, p. 1004-1013.

Crawford, M. L. (1981) Phase equilibria in aqueous fluid inclusions: in Hollister, L. S. and Crawford M. L., eds., Short course in fluid inclusions: Applications to petrology. Mineralogical Assoc. Canada, p. 75-100.

Czerniakowski, L. A., Lohmann, K. C., and Wilson, J. L. (1984) Closed-system marine burial diagenesis - isotopic chemistry of Austin Chalk: Sedimentology, Vol. 31, p. 863-877.

Davies, G. R. and Krouse, H. R. (1975) Carbon and oxygen isotopic composition of Late Paleozoic calcite cements, Canadian Arctic archipelago - preliminary results and interpretation: Geol. Soc. Canada, Paper 75-1, Part B, p. 215-220.

Degens, E. T. and Epstein, S. (1964) Oxygen and carbon isotope ratios in coexisting calcites and dolomites from recent and ancient sediments: Geochim. Cosmochim. Acta, Vol. 28, p. 23-44.

DePaolo, D. J. (1986) Detailed record of the Neogene strontium isotopic evolution of seawater from DSDP site 5908: Geology, Vol. 14, p. 103-106.

Dickson, J. A. D. (1966) Carbonate identification and genesis as revealed by staining: Jour. Sed. Petrology, Vol. 36, p. 491-505.

Dravis, J. J. and Yurewicz, D. A. (1985) Enhanced carbonate petrography using fluorescence microscopy: Jour. Sed. Petrology, Vol. 55, p. 795-804.

Dunham, J. B. and Olson, E. R. (1980) Shallow subsurface dolomitization of subtidally deposited carbonate sediments in the Hanson Creek Formation (Ordovician - Silurian) of Central Nevada: in Zenger, D. H., Dunham, J.B., and Ethington, R. L., eds., Concepts and Models of Dolomitization, Soc. Econ. Paleontologists and Mineralogists Spec. Publ. 28, p. 139-161.

Dunn, P. A. (1988) Dyanmics of $\delta^{13}C$ and $\delta^{18}O$ variation in the Devono-Carboniferous succession of Belgium and Ireland: unpubl. M.Sc. thesis, Univ. of Michigan, Ann Arbor, MI, 55 p.

Elderfield, H. (1986) Strontium isotope stratigraphy: in Shackleton, N.J., ed., Boundaries and Events in the Paleogene, Palaeogeography, Palaeoclimatology, Palaeoecology, Vol. 57, p. 71-90.

Emrich, K., Ehhalt, D. H., and Vogel, J. C. (1970) Carbon isotope fractionation during the precipitation of calcium carbonate: Earth Planet. Sci. Lett., Vol. 8, p. 363-371.

Engel, A. E. J., Clayton, R. N., and Epstein, S. (1958) Variations in isotopic composition of oxygen and carbon in Leadville limestone (Mississippian, Colorado) and in its hydrothermal and metamorphic phases: Jour. Geol., Vol. 66, p. 374-393.

Enos, P. (1988) Evolution of pore space in the Poza Rica trend (Mid-Cretaceous), Mexico: Sedimentology, Vol. 35, p. 287-325.

Exploration Staff, Chevron Standard Limited (1979) The geology, geophysics, and significance of the Nisku Reef discoveries, West Pembina Area, Alberta, Canada: Bull. Can. Petrol. Geol., Vol. 27, p. 326-359.

Faure, G. (1977) Principles of Isotope Geology: John Wiley and Sons, New York, 464 p.

Faure, G. and Powell, J. L. (1972) Strontium Isotope Geology: Springer Verlag, Berlin, 18 p.

Feazel, C. T. (1985) Diagenesis of Jurassic grainstone reservoirs in the Smackover Formation, Chatom Field, Alabama: in Roehl, P. O. and Choquette, P. W., eds., Carbonate Petroleum Reservoirs, Springer-Verlag, New York, p. 359-367.

Frank, J. R., Carpenter, A. B., and Oglesby, T. W. (1982) Cathodoluminescence and composition of calcite cement in the Taum Sauk Limestone (Upper Cambrian), Southeast Missouri: Jour. Sed. Petrology, Vol. 52, p. 631-638.

Friedman, G. M. (1989) Case history of deep-burial sulfide mineralization in the northern Appalachian Basin: Carbonates and Evaporites, Vol. 4, p. 231-241.

Friedman, I. and O'Neil, J. R. (1977) Compilation of stable isotope fractionation factors of geochemical interest: USGS Professional Paper 440KK, 12 p.

Fritz, P. (1967) Oxygen and carbon isotopic composition of carbonates from the Jura of southern Germany: Canadian Jour. Earth Sci., Vol. 4, p. 1247-1267.

Fritz, P. (1969) The oxygen and carbon isotopic composition of carbonates from the Pine Point lead-zinc ore deposits: Econ. Geol., Vol. 64, p. 733-742.

Fritz, P. and Jackson, S. A. (1972) Geochemical and isotopic characteristics of Middle Devonian dolomites from Pine Point, northern Canada: Proc. 24th Int'l. Geol. Congress, Section 6, p. 230-243.

Gao, G. (1990) Geochemical and isotopic constraints on the diagenetic history of a massive stratal, late Cambrian (Royer) dolomite, Lower Arbuckle Group, Slick Hills, southwestern Oklahoma, USA: Geochim. Cosmochim. Acta, Vol. 54, p. 1979-1989.

Gao, G. and Land, L. S. (1991) Early Ordovician Cool Creek Dolomite, Middle Arbuckle Group, Slick Hills, SW Oklahoma, U.S.A.: origin and modification: Jour. Sed. Petrology, Vol. 61, p. 161-173.

Garber, R. A., Harris, P. M., and Borer, J. M. (1990) Occurrence and significance of magnesite in Upper Permian (Guadalupian) Tansill and Yates Formations, Delaware Basin, New Mexico: Amer. Assoc. Petroleum Geologists Bull., Vol. 74, p. 119-134.

Garven, G. and Freeze, R. A. (1984) Theoretical analysis of the role of groundwater flow in the genesis of stratabound ore deposits - I. Mathematical and numerical model: Amer. Jour. of Sci., Vol. 284, p.1085-1124.

Gill, D. (1985) Depositional facies of Middle Silurian (Niagaran) pinnacle reefs, Belle River Mills Gas Field, Michigan Basin, southeastern Michigan: in Roehl, P. O. and Choquette, P. W., eds., Carbonate Petroleum Reservoirs, Springer-Verlag, New York, p. 121-139.

Given, R. K., and Lohmann, K. C. (1985) Derivation of the original isotopic composition of Permian marine carbonates: Jour. Sed. Petrology, Vol. 55, p. 430-439.

Goldstein, R. H. (1986) Integrative carbonate diagenesis studies: Fluid inclusions in calcium carbonate cement; paleosols and cement stratigraphy of Late Pennsylvanian cyclic strata, New Mexico: Ph.D. Diss., Univ. of Wisconsin, Madison, 340 pages.

Goldstein, R. H. (1986) Re-equilibration of fluid inclusions in low-temperature calcium-carbonate cement: Geology, V. 14, p. 792-795.

Goldstein, R. H. and Reynolds, T. J. (in press) Fluid inclusions in diagenetic minerals: Soc. Econ. Paleontologist and Mineralogists, Short Course Notes.

Gonzales, L. A. and Lohmann, K. C. (1985) Carbon and oxygen isotopic composition of Holocene reefal carbonate: Geology, Vol. 13, p. 811-814.

Graber, E. R. and Lohmann, K. C. (1989) Basinal marine dolomite from the Pennsylvanian Wolfcampian Horquilla Formation, New Mexico: Jour. Sed. Petrology, Vol. 59, p. 4-12.

Gregg, J. M. (1985) Regional epigenetic dolomitization in the Bonneterre Dolomite (Cambrian), southeastern Missouri: Geology, Vol. 13, p. 503-506.

Grossman, E. L., Yancey, T. E., Zhang, C., and Mii, H.-S. (1990) Carbon and oxygen isotope variations in Pennsylvanian brachiopods from North America: Geol. Soc. Amer., Abs. with Prog., p. A115 (abs.).

Gruszczynski, M., Halas, S., Hoffman, A., and Malkowski, K. (1989) A brachiopod calcite record of the oceanic carbon and oxygen isotope shifts at the Permian/Triassic transition: Nature, Vol. 337, p. 64-68.

Hall, W. E., and Friedman, I. (1969) Oxygen and carbon isotopic composition of ore and host rock of selected Mississippi Valley deposits: USGS Prof. Paper 650-C, p. C140-C148.

Hanor, J. S. (1987) Origin and migration of subsurface sedimentary brines: Soc. Econ. Paleontologists and Mineralogists Short Course No. 21, 247 p.

Hardie, L. A. (1987) Dolomitization: a critical view of some current views: Jour. Sed. Petrology, Vol. 57, p. 166-183.

Harwood, G. and Fontana, C. (1984) Smackover deposition and diagenesis and structural history of the Bryan's Mill area, Cass and Bowie counties, Texas: in Ventress, W.P.S., Bebout, D.G., Perkins, B.F., and Moore, C.H., eds., The Jurassic of the Gulf Rim, Proc. Third Ann. Rsh. Conf., Gulf Coast Sect., Soc. Econ. Paleontologists and Mineralogists, p. 135-147.

Haynes, F. M. (1985) Determination of fluid inclusion compositions by sequential freezing: Econ. Geol., Vol. 80, p. 1436-1439.

Haynes, F. M. (1986) Geologic and geochemical controls for sphalerite mineralization, Mascot-Jefferson city zinc district, east Tennessee: unpubl. Ph. D. dissert., Univ. of Michigan, Ann Arbor, MI, 232 p.

Hemming, N. G., Meyers, W. J., and Grams, J. C. (1989) Cathodoluminescence in diagenetic calcites: the roles of Fe and Mn as deduced from electron probe and spectrophotometric measurements: Jour. Sed. Petrology, Vol. 59, p. 404-411.

Heydari, E. and Moore, C. H. (1989) Burial diagenesis and thermochemical sulfate reduction, Smackover Formation, southeastern Mississippi salt basin: Geology, Vol. 17, p. 1080-1084.

Hild, G. P. (1986) The relationship of San Andres facies to the distribution of porosity and permeability - Garza Field, Garza County, Texas: in Bebout, D. G. and Harris, P. M., eds., Hydrocarbon Reservoir Studies, San Andres/Grayburg Formations, Permian Basin, PBS-SEPM Publ. No. 86-26, p. 17-20.

Hitchon, B. and Friedman, I (1969) Geochemistry and origin of formation waters in the western Canada sedimentary basin- I. Stable isotopes of hydrogen and oxygen: Geochim. Cosmochim. Acta, Vol. 33, p. 1321-1349.

Hoefs, J. (1980) Stable Isotope Geochemistry (2nd ed.): Springer Verlag, Berlin, 208 p.

Holail, H., Lohmann, K. C., and Sanderson, I. (1988) Dolomitization and dedolomitization of Upper Cretaceous carbonates: Bahariya Oasis, Egypt: in Shukla, V. and Baker, P. A., eds., Sedimentology and Geochemistry of Dolostones, Soc. Econ. Paleontologists and Mineralogists Spec. Publ. 43, p. 235-243.

Hudson, J. D. (1977) Stable isotope studies and limestone lithification: Jour. Geol. Soc. London, Vol. 133, p. 637-660.

Humphrey, J. D. (1988) Late Pleistocene mixing zone dolomitization, southeastern Barbados, West Indies: Sedimentology, Vol. 35, p. 327-348.

Hurley, N. F., and Lohmann, K. C. (1989) Diagenesis of Devonian reefal carbonates in the Oscar Range, Canning Basin, western Australia: Jour. Sed. Petrology, Vol. 59, P. 127-146.

Illing, L. V. (1959) Deposition and diagenesis of some Upper Palaeozoic carbonate sediments in Western Canada: in Proc. Fifth World Petrol. Congr., Sect. I, Paper 2, p. 23-50.

Irwin, H., Curtis, M. L. and Coleman, C. (1977) Isotopic evidence for the source of diagenetic carbonate during burial of organic-rich sediments: Nature, Vol. 269, p. 209-213.

James, A. T. and Burns, B. J. (1984) Microbial alteration of subsurface natural gas accumulations: Amer. Assoc. Petroleum Geologists Bull., Vol. 68, p. 957-960.

James, N. P. and Choquette, P. W. (1983) Diagenesis 6: Limestones - the seafloor diagenetic environment: Geosci. Canada, Vol. 10, p. 162-179.

James, N. P. and Ginsburg, R. N. (1979) The Seaward Margin of Belize Barrier and Atoll Reefs: Intl. Assoc. Sedimentologists Spec. Publ. 3, 191 p.

James, N. P. and Klappa, C. F. (1983) Petrogenesis of Early Cambrian reef limestones, Labrador, Canada: Jour. Sed. Petrology, Vol. 53, p. 1051-1096.

Kaufman, A. and Magaritz, M. (1980) The climatic history of the eastern Mediterranean as recorded in mollusk shells: Radiocarbon, Vol. 22, p. 778-781.

Kaufman, J. (1989) Sedimentology and diagenesis of the Swan Hills Formation (Middle-Upper Devonian) Rosevear Field: unpubl. Ph. D. dissert., State Univ. of New York, Stony Brook, NY, 412 pp.

Kharaka, Y. H., Carothers, W. W. and Rosenbauer, R. J. (1983) Thermal decarboxylation of acetic acid: Implications for origin of natural gas: Geochim. Cosmochim. Acta, Vol. 47, p. 397-402.

Kier, J. S. (1973) Primary subtidal dolomite from Baffin Bay, Texas: Amer. Assoc. Petroleum Geologists Bull., Vol. 57, p. 788.

Kinsman, D. J. J. and Patterson, R. J. (1973) Dolomitization process in sabkha environment: Amer. Assoc. Petroleum Geologists Bull., Vol. 57, p. 788-789.

Knauth, L. P. and Beeunas, M. A. (1986) Isotope geochemistry of fluid inclusions in Permian halite with implications for the isotopic history of ocean water and the origin of saline formation waters: Geochim. Cosmochim. Acta, Vol. 50, p. 419-433.

Kohout, F. A. (1967) Groundwater flow and the geothermal regime of the Floridian plateau: Trans. Gulf Coast Assoc. Geol. Soc., Vol. 17, p. 339-354.

Krouse, H. R., Viau, C. A., Eliuk, L. S., Ueda, A., and Halas, S. (1988) Chemical and isotopic evidence of thermochemical sulphate reduction by light hydrocarbon gases in deep carbonate reservoirs: Nature, Vol. 333, p. 415-419.

Land, L. S. (1973a) Holocene meteoric dolomitization of Pleistocene limestones, North Jamaica: Sedimentology, Vol. 20, p. 411-424.

Land, L. S. (1973b) Contemporaneous dolomitization of Middle Pleistocene reefs by meteoric water, North Jamaica: Bull. Mar. Sci., Vol. 23, p. 64-92.

Land, L. S. (1980) The isotopic and trace element geochemistry of dolomite: the state of the art: in Zenger, D. H., Dunham, J. B., and Ethington, R. L., Concepts and Models of Dolomitization, Soc. Econ. Paleontologists and Mineralogists Spec. Publ. 28, p. 87-110.

Land, L. S. (1986) Environments of limestone and dolomite diagenesis: some geochemical considerations: in Warme, J. E. and Shanley, K. W., eds., Carbonate Depositional Environments, Modern and Ancient: Part 5: Diagenesis 1, Colorado School of Mines Quart., Vol. 81, p.26-41.

Land, L. S. (1989) Dolomitization and dolomite recrystallization, Hope Gate Formation, North Jamaica: reassessment of mixing-zone dolomite: Geol. Soc. Amer., Abs. with Prog., p. A220-A221, (abs.).

Land, L. S. and Dutton, S. P. (1978) Cementation of a Pennsylvanian deltaic sandstone: isotopic data, Jour. Sed. Petrology, Vol. 48, p. 1167-1176.

Land, L. S. and Epstein, S. (1970) Late Pleistocene diagenesis and dolomitization, North Jamaica: Sedimentology, Vol. 14, p. 187-200.

Land, L. S. and Fisher, R. S. (1987) Wilcox sandstone diagenesis, Texas Gulf Coast: a regional isotopic comparison with the Frio Formation: in J. Marshall, ed., Diagenesis of Sedimentary Sequences, Geological Soc. Spec. Publ. 36, p. 219-235.

Land, L. S. and Prezbindowski, D. R. (1981) The origin and evolution of saline formation water, Lower Cretaceous carbonates, South-central Texas: Jour. of Hydrology, Vol. 54, p. 51-74.

Land, L. S. and Prezbindowski, D. R. (1985) Chemical constraints and origins of four groups of Gulf Coast reservoir fluids: Discussion: Amer. Assoc. Petroleum Geologists Bull., Vol. 69, p. 119-126.

Land, L. S., Milliken, K. L., and McBride, E. F. (1987) Diagenetic evolution of Cenozoic sandstones, Gulf of Mexico sedimentary basin: Sed. Geology, Vol. 50, p. 195-225.

Land, L. S., Salem, M R. I., and Morrow, D. W. (1975) Paleohydrology of ancient dolomites: geochemical evidence: Amer. Assoc. Petroleum Geologists Bull., Vol. 59,, p. 1602-1625.

Leach, D. L. and Rowen, E. L. (1986) Genetic link between Ouachita foldbelt tectonism and the Mississippi Valley-type lead-zinc deposits of the Ozarks: Geology, Vol. 14, p. 931-935.

Leary, D. A., and Vogt, J. N. (1986) Diagenesis of the Permian (Guadalupian) San Andres Formation, Central Basin Platform, West Texas: in Bebout, D. G. and Harris, P. M., eds., Hydrocarbon Reservoir Studies, San Andres/Grayburg Formations, Permian Basin, PBS-SEPM Publ. 86-26, p.67-68.

Lee, Y. I. and Friedman, G. M. (1987) Deep-burial dolomitization in the Ordovician Ellenburger Group carbonates, West Texas and southeastern New Mexico: Jour. Sed. Petrology, Vol. 57, p. 544-557.

Lindsay, R. F. and Kendall, C. G. St. C. (1985) Depositional facies, diagenesis and reservoir character of Mississippian cyclic carbonates in the Mission Canyon Formation, Little Knife Field, Williston Basin, North Dakota: in Roehl, P. O. and Choquette, P. W., eds., Carbonate Petroleum Reservoirs, Springer-Verlag, New York, p. 177-190.

Lloyd, R. M. (1969) A paleoecological interpretation of the Caloosahatchee Formation, using stable isotope methods: Jour. Geol., Vol. 77, p. 1-25.

Lohmann, K. C. (1988a) Application of carbon and oxygen isotopic techniques for unravelling the diagenetic history of carbonate sequences: in Allan, J. R. and Harris, P. M., eds., Stable Isotope, Trace Element, and Fluid Inclusion Workshop, Chevron Oil Field Research Co., unpublished report, p. 1-49.

Lohmann, K. C. (1988b) Geochemical patterns of meteoric diagenetic systems and their application to studies of paleokarst: in James, N. P. and Choquette, P. W., eds., Paleokarst, Springer-Verlag, New York, p. 58-80.

Lohmann, K. C., and Walker, J. C. G. (1989) The $\delta^{18}O$ record of Phanerozoic abiotic marine calcite cements: Geophys. Rsh. Letters, Vol. 16, p. 319-322.

Long, K. R., Kelly, W. C., Ohle, E. L., and Lohmann, K. C. (1986) Ground preparation and zinc mineralization in bedded and breccia ores of the Monte Cristo mine, North Arkansas: Econ. Geol., Vol. 81, p. 809-830.

Lowenstam, H. A. (1961) Mineralogy, O^{18}/O^{16} ratios, and strontium and magnesium contents of recent and fossil brachiopods and their bearing on the history of the oceans: Jour. Geol., Vol. 69, p. 241-260.

Lowenstam, H. A. and Epstein, S. (1954) Paleotemperatures of the post-Aptian Cretaceous as determined by the oxygen isotope method: Jour. Geol., Vol. 62, p. 207-248.

Ludvigson, G. A., Witzke, B. J., Plocher, O. W., Gonzalez, L. A., and Lohmann, K. C. (1990) Secular variation in carbon - oxygen isotopic composition of Middle Ordovician marine carbonate: Geol. Soc. Amer., Abs. with Prog., p. A115 (abs.).

M'Rabet, A. (1981) Differentiation of environments of dolomite formation, Lower Cretaceous of Central Tunisia: Sedimentology, Vol. 28, p. 331-352.

Machel, H. G. (1985) Facies and diagenesis of the Upper Devonian Nisku Formation in the subsurface of Central Alberta: unpubl. Ph.D. dissert., McGill Univ., Montreal, Canada, 392 pp.

Machel, H. G. (1986) Early lithification, dolomitization and anhydritization of Upper Devonian Nisku buildups, subsurface of Alberta, Canada: in Schroeder, J. H. and Purser, B. H., eds., Reef Diagenesis, Springer-Verlag, Berlin, p. 336-356.

Machel, H. G. and Anderson, J. H. (1989) Pervasive subsurface dolomitization of the Nisku Formation in Central Alberta: Jour. Sed. Petrology, Vol. 59, p. 891-911.

Machel, H. G. and Burton, E. A. (1991) Factors governing cathodoluminescence in calcite and dolomite, and their implications for studies of carbonate diagenesis: in Barker, C. E. and Kopp, O. C., eds., Luminescence Microscopy and Spectroscopy: Qualitative and Quantitative Applications, Soc. Econ. Paleontologists and Mineralogists, Short Course Notes, No. 25, p. 1-7.

Magaritz, M., Goldenberg, L., Kafri, U. and Arad, A. (1980) Dolomite formation in the seawater-freshwater interface: Nature, Vol. 287, p. 622-624.

Marshall, J. D. and Ashton, M. (1980) Isotopic and trace element evidence for submarine lithification of hardgrounds in the Jurassic of eastern England: Sedimentology, Vol. 27, p. 271-289.

Marshall, J. D. and Middleton, (1990) Changes in marine isotopic composition and the late Ordovician glaciation: Jour. Geol. Soc. London, Vol. 147, p. 1-4.

Mattes, B. W. and Mountjoy, E. W. (1980) Burial dolomitization of the upper Devonian Miette buildup, Jasper National Park, Alberta: in Zenger, D. H., Dunham, J. B., and Ethington, R. L., eds., Concepts and Models of Dolomitization, Soc. Econ. Paleontologists and Mineralogists Spec. Publ. 28, p. 259-297.

Mazzullo, S. J., Reid, A. M. and Gregg, J. M. (1987) Dolomitization of Holocene Mg-calcite supratidal deposits, Ambergris Cay, Belize: Bull. Geol. Soc. Amer., Vol. 98, p. 224-231.

McKenzie, J. A. (1981) Holocene dolomitization of calcium carbonate sediments from the coastal sabkhas of Abu Dhabi, U.A.E.: a stable isotope study: Jour. Geol., Vol. 89, p. 185-198.

McKenzie, J. A., Hsu, K. J., and Schneider, J. F. (1980) Movement of subsurface waters under the sabkha, Abu Dhabi, UAE, and its relation to evaporative dolomite genesis: in Zenger, D. H., Dunham, J. B., and Ethington, R.L., eds., Concepts and Models of Dolomitization, Soc. Econ. Paleontologists and Mineralogists Spec. Publ. 28, p. 11-20.

McLimans, R. K. (1987) The application of fluid inclusions to migration of oil and diagenesis in petroleum reservoirs: Applied Geochem., Vol. 2, p. 585-603.

Meyers, W. J., and Lohmann, K. C. (1985) Isotope geochemistry of regionally extensive calcite cement zones and marine components in Mississippian limestones, New Mexico: in Harris, P. M. and Schneidermann, N., eds., Carbonate Cements, Soc. Econ. Paleontologists and Mineralogists Spec. Publ. No. 36, p. 223-240.

Miller, K. G., Fairbanks, R. G., and Mountain, G. S. (1987) Tertiary oxygen isotope synthesis, sea level history, and continental margin erosion: Paleoceanography, Vol. 2, p. 1-19.

Mitchell, J. T., Land, L. S., and Miser, D. E. (1987) Modern marine dolomite cement in a North Jamaican fringing reef: Geology, Vol. 15, p 557-560.

Moldovanyi, E. P. and Lohmann, K. C. (1984) Isotopic criteria for recognition of successive events of phreatic cementation, Sligo and Cupido Formations: Jour. Sed. Petrology, Vol. 54, p. 972-985.

Moore, C. H. (1989) Carbonate Diagenesis and Porosity: Developments in Sedimentology No. 46, Elsevier, New York, 338 pp.

Moore, C. H., Chowdhury, A., and Chan, L. (1988) Upper Jurassic Smackover platform dolomitization, northwestern Gulf of Mexico: a tale of two waters: in Shukla, V. and Baker, P. A., eds., Sedimentology and Geochemistry of Dolostones, Soc. Econ. Paleontologists and Mineralogists Spec. Publ. 43, p.175-189.

Morrison, J. O., and Brand, U. (1988) An evaluation of diagenesis and chemostratigraphy of Upper Cretaceous molluscs from the Canadian interior seaway: Chem. Geol., Vol. 72, p. 235-248.

Morrison, J. O., and Veizer, J. (1990) Belemnite geochemistry: an indicator of secular and environmental variation of Cretaceous seawater: Geol. Soc. Amer., Abs. with Prog., p. A116.

Morrow, D. W. (1982a) Diagenesis 1. Dolomite - Part 1: The chemistry of dolomitization and dolomite precipitation: Geosci. Canada, Vol. 9, p. 5-13.

Morrow, D. W. (1982b) Diagenesis 2. Dolomite - Part 2: Dolomitization models and and ancient dolostones: Geosci. Canada, Vol. 9, p. 95-107.

Morrow, D. W., Cumming, G. L., and Koepnick, R. B. (1986) Manetoe Facies - a gas-bearing, megacrystalline, Devonian dolomite, Yukon and Northwest Territories, Canada: Amer. Assoc. Petroleum Geologists Bull., Vol. 70, p. 702-720.

Morse, J. W. and Mackenzie, F. T. (1990) Geochemistry of Sedimentary Carbonates: Elsevier, New York, 707p.

Mountjoy, E. W., and Walls, R. A. (1977) Some examples of early submarine cements from Devonian buildups of Alberta: in Proceedings, Third International Coral Reef Symposium, Miami, FL, p. 155.

Mullins, H. T., Dix, G. R., Garduski, A. F., and Land, L. S. (1988) Neogene deep-water dolomite from the Florida-Bahamas platform: in, Shukla, V. and Baker, P. A., eds., Sedimentology and Geochemistry of Dolostones, Soc. Econ. Paleontologists and Mineralogists Spec. Publ. 43, p. 235-243.

O'Shea, K. J., Miles, M. C., Fritz, P., Frape, S. K., and Lawson, D. E. (1988) Oxygen-18 and carbon-13 in the carbonates of the Salina Formation of southwestern Ontario: Can. Jour. Earth Sci., Vol. 25, p. 182-194.

Oliver, J. (1986) Fluids expelled tectonically from orogenic belts: their role in hydrocarbon migration and other geologic phenomena: Geology, Vol. 14, p.99-102.

Pedone, V. (1990) Regional dolomitization of Devonian reef complexes, Canning Basin, Western Australia: unpubl. Ph.D. dissert., State Univ. of New York, Stony Brook, N. Y.

Peryt, T. M. and Magaritz, M. (1990) Genesis of evaporite-associated platform dolomites: case study of the Main Dolomite (Zechstein, Upper Permian), Leba elevation, northern Poland: Sedimentology, Vol. 37, p. 745-761.

Pierre, C., Ortlieb, L., Person, A. (1984) Supratidal evaporitic dolomite at Ojo de Liebre lagoon: mineralogical and isotopic arguments for primary crystallization: Jour. Sed. Petrology, Vol. 54, p. 1049-1061.

Pierson, B. J. (1982) Cyclic sedimentation, limestone diagenesis and dolomitization in upper Cenozoic carbonates of the southeastern Bahamas: unpubl. Ph.D. dissert., Univ. Miami, Miami, FL.

Pisciotto, K. A. and Mahoney, J. J. (1981) Isotopic survey of diagenetic carbonates, Deep Sea Drilling Project Leg 63: in Orlofsky, S., ed., Initial Reports of the Deep Sea Drilling Project, Vol. 63, p. 595-609.

Popp, B. N., Anderson, T. F. and Sandberg, P. A. (1986a) Brachiopods as indicators of original isotopic compositions in some Paleozoic limestones: Bull. Geol. Soc. Amer., Vol. 97, p. 1262-1269.

Popp, B. N., Anderson, T. F. and Sandberg, P. A. (1986b) Textural, elemental, and isotopic variations among constituents in Middle Devonian limestones, North America: Jour. Sed. Petrology, Vol. 56, p. 715-727.

Qing, H. and Mountjoy, E. W. (1989) Multistage dolomitization in Rainbow buildups, Middle Devonian Keg River Formation, Alberta, Canada: Jour. Sed. Petrology, Vol. 59, p. 114-126.

Radke, B. M. and Mathis, R. L. (1980) On the formation and occurrence of saddle dolomite: Jour. Sed. Petrology, Vol. 50, p. 1149-1168.

Randazzo, A. F. and Cook, D. J. (1987) Characterization of dolomitic rocks from the coastal mixing zone of the Floridan aquifer, Florida, U.S.A.: Sed. Geology, Vol. 54, p. 169-192.

Randazzo, A. F. and Zachos, L.G. (1983) Classification and description of dolomitic fabrics of rocks from the Floridan aquifer, U.S.A.: Sed. Geology, Vol. 37, p. 151-162.

Reeder, R. J. (1981) Electron optical investigation of sedimentary dolomites: Contrib. Miner. Petrol., Vol. 76, P. 148-157.

Riecker, R. E. (1962) Hydrocarbon fluorescence and migration of petroleum: Amer. Assoc. Petroleum Geologists Bull., Vol. 46, p. 60-75.

Rittenhouse, G. (1967) Bromine in oil-field waters and its use in determining possibilities of origin of these waters: Amer. Assoc. Petroleum Geologists Bull., Vol. 51, p. 2430-2440.

Roedder, E. (1981) Origin of fluid inclusions and changes that occur after trapping: in Hollister, L. S. and Crawford M. L., eds., Short Course in Fluid Inclusions: Applications to Petrology. Mineralogical Assoc. Canada, p. 101-137.

Roehl, P. O. (1985) Depositional and diagenetic controls on reservoir rock development and petrophysics in Silurian tidalites, Interlake Formation, Cabin Creek Field area, Montana: in Roehl, P. O. and Choquette, P. W., eds., Carbonate Petroleum Reservoirs, Springer-Verlag, New York, p. 87-105.

Rosen, M. R., Miser, D. E. and Warren, J. K. (1988) Sedimentology, mineralogy, and isotopic analysis of Pellet Lake, Coorong region, South Australia: Sedimentology, Vol. 35, p. 105-122.

Ruppel, S. C. (1986) San Andres facies and porosity distribution - Emma Field, Andrews County, Texas: in Bebout, D. G. and Harris, P. M., eds., Hydrocarbon Reservoir Studies, San Andres/ Grayburg Formations, Permian Basin, PBS-SEPM Publ. No. 86-26, p.99-103.

Ruppel, S. C., and Cander, H. S. (1988) Dolomitization of shallow-water platform carbonates by sea water and seawater-derived brines: San Andres Formation (Guadalupian), West Texas: in Shukla, V. and Baker, P. A., eds., Sedimentology and Geochemistry of Dolostones, Soc. Econ. Paleontologists and Mineralogists Spec. Publ. 43, p. 245-262.

Ruzyla, K., and Friedman, G. M. (1985) Factors controlling porosity in dolomite reservoirs of the Ordovician Red River Formation, Cabin Creek Field, Montana: in Roehl, P. O. and Choquette, P. W., eds., Carbonate Petroleum Reservoirs, Springer-Verlag, New York, p. 41-58.

Saller, A. H. (1984) Petrologic and geochemical constraints on the origin of subsurface dolomite, Enewetak Atoll: an example of dolomitization by normal seawater: Geology, Vol. 12, p. 217-220.

Sassen, R. (1988) Geochemical and carbon isotopic studies of crude oil destruction, bitumen precipitation, and sulfate reduction in the deep Smackover Formation: Organic Geochemistry, Vol. 12, p. 351-361.

Savin, S. M., Douglas, R. G., and Stehli, F. G. (1975) Tertiary marine paleotemperatures: Bull. Geol. Soc. Amer., Vol. 86, p. 1499-1510.

Schmoker, J. W. and Halley, R. B. (1982) Carbonate porosity versus depth: a predictable relation for South Florida: Amer. Assoc. Petroleum Geologists Bull., Vol. 66, p. 2561-2570.

Schultz, J. L., Boles, J. R., and Tilton, G. R. (1989) Tracking calcium in the San Joaquin Basin, California: A strontium isotope study of carbonate cements at North Coles Levee: Geochim. Cosmochim. Acta, Vol. 53, p. 1991-1999.

Sears, S. O. and Lucia, F. J. (1980) Dolomitization of northern Michigan Niagara reefs by brine refluxion and freshwater/seawater mixing: in Zenger, D. H., Dunham, J. B., and Ethington, R. L., eds., Concepts and Models of Dolomitization, Soc. Econ. Paleontologists and Mineralogists Spec. Publ. 28, p.215-235.

Sellwood, B. W., and Shepherd, T. J., and Evans, M. R., and James, B. (1989) Origin of late cements in oolitic reservoir facies: a fluid inclusion and isotopic study (Mid-Jurassic, southern England): Sed. Geology, Vol. 61, p. 223-237.

Shepherd, T., Rankin, A. H., and Alderton, D. H. M. (1985) A Practical Guide to Fluid Inclusion Studies: Blackie and Son Ltd., New York, 239pp.

Shinn, E. A. (1983) Tidal flat environment: in Scholle, P. A., Bebout, D. G., and Moore, C. H., eds., Carbonate Depositional Environments, Amer. Assoc. Petroleum Geologists Memoir 33, p. 172-210.

Sibley, D. F. (1980) Climatic control of dolomitization, Seroe Domi Formation (Pliocene), Bonaire, N. A.: in Zenger, D. H., Dunham, J. B., and Ethington, R. L., eds., Concepts and Models of Dolomitization, Soc. Econ. Paleontologists and Mineralogists Spec. Publ. 28, p. 247-258.

Sibley, D. F. (1991) Secular changes in the amount and texture of dolomite: Geology, Vol. 19, p. 151-154.

Spencer, R. J. (1987) Origin of Ca-Cl brines in Devonian formations, western Canada sedimentary basin: Applied Geochem., Vol. 2, p. 373-384.

Stepanek, B. E. (1987) Dolomitization of paleokarst collapse breccias - the Ordovician El Paso Group, Franklin Mountains, West Texas: in Field Trip Guidebook, Mega-collapse Breccia and Associated late Stage Dolomitization of Ordovician Carbonates Franklin Mountains, West Texas, Soc. Econ. Paleontologists and Mineralogists Midyear Meeting.

Sternbach, C. A. and Friedman, G. F. (1986) Dolomites formed under conditions of deep burial: Hunton Group carbonate rocks (Upper Ordovician to Lower Devonian) in the the deep Anadarko Basin of Oklahoma and Texas: Carbonates and Evaporites, Vol. 1, p. 61-73.

Stoakes, F. A. (1987) Fault-controlled dolomitization of the Wabamun Group, Tangent Field, Peace River Arch, Alberta: in Krause, F. F. and Burrows, O. G., eds., Devonian Lithofacies and Reservoir Styles in Alberta, 13th Can. Soc. Petrol. Geol. Core Conf. and Display, p. 73-85.

Stoakes, F. A., and Wendte, J. C. (1987) The Woodbend Group: in Krouse, K. K. and Burrowes, O. G., eds., Devonian Lithofacies and Reservoir Styles in Alberta, Can. Soc. Petrol. Geol. Core Conf. No. 13, p. 153-170.

Supko, P. R. (1977) Subsurface dolomites, San Salvador, Bahamas: Jour. Sed. Petrology, Vol. 47, p. 1963-1077.

Swart, P. K., Ruiz, J., and Holmes, C. W. (1987) Use of strontium isotopes to constrain the timing and mode of dolomitization of upper Cenozoic sediments in a core from San Salvador, Bahamas: Geology, Vol. 15, p. 262-265.

Tan, F. C. and Hudson, J. D. (1971) Carbon and oxygen isotopic relationships of dolomites and co-existing calcites, Great Estuarine Series (Jurassic), Scotland: Geochim. Cosmochim. Acta, Vol. 35, p. 755-767.

Tan, F. C. and Hudson, J. D. (1974) Isotopic studies on the palaeoecology and diagenesis of the Great Estuarine Series (Jurassic) of Scotland: Scottish Jour. of Geology, Vol. 10, p. 91-128.

Tan, F. C., Hudson, J. D., and Keith, M. L. (1970) Jurassic (Callovian) paleotemperatures from Scotland: Earth Planet. Sci. Letters, Vol. 9, p. 421-426.

Taylor, H. P. (1974) The application of oxygen and hydrogen isotope studies to problems of hydrothermal alteration and ore deposition: Econ. Geol., Vol. 69, p. 843-883.

Trudinger, P. A., Chambers, L. A., and Smith, J. W. (1985) Low-temperature sulphate reduction: Biological versus abiological: Can. Jour. Earth Sci., Vol. 22, p. 1910-1918.

Veizer, J. (1970) Zonal arrangement of the Triassic rocks of the Western Carpathians: a contribution to the dolomite problem: Jour. Sed. Petrology, Vol. 40, p. 1287-1301.

Veizer, J. (1983) Chemical diagenesis of carbonates: Theory and application of trace element technique: in Stable Isotopes in Sedimentary Geology, Soc. Economic Paleontologists and Mineralogists Short Course No. 10, p. 3-1 to 3-100.

Veizer, J. (1989) Strontium isotopes in seawater through time: Ann. Rev. Earth Planet. Sci. Letters, Vol. 17, p. 141-167.

Veizer, J. and Compston, W. (1974) $^{87}Sr/^{86}Sr$ composition of seawater during the Phanenozoic: Geochim. Cosmochim. Acta, Vol. 38, p. 1461-1484.

Veizer, J. and Fritz, P. (1976) Possible control of post-depositional alteration in oxygen paleotemperature determinations: Earth Planet. Sci. Letters, Vol. 33, p. 255-260.

Veizer, J. and Hoefs, J. (1976) Nature of O^{18}/O^{16} and C^{13}/C^{12} secular trends in sedimentary carbonate rocks: Geochim. Cosmochim. Acta, Vol. 40, p. 1387-1395.

Veizer, J., Fritz, P., and Jones, B. (1986) Geochemistry of brachiopods: oxygen and carbon isotopic records of Paleozoic oceans: Geochim. Cosmochim. Acta, Vol. 50, p. 1679-1696.

Viau, C. A., and Oldershaw, A. E. (1984) Structural controls on sedimentation and dolomite cementation in the Swan Hills Formation, Swan Hills Field, central Alberta: in Eliuk, L., ed., Carbonates in Subsurface and Outcrop , Can. Soc. Petrol. Geol. Core Conf., p. 103-131.

Ward, R. F., Kendall, C. G. St.C, and Harris, P. M. (1986) Upper Permian (Guadalupian) facies and their association with hydrocarbons - Permian Basin, West Texas and New Mexico: Amer. Assoc. Petroleum Geologists Bull., Vol. 70, p. 239-262.

Ward, W. C., and Halley, R. B. (1985) Dolomitization in a mixing zone of near-seawater composition, Late Pleistocene, northeastern Yucatan Peninsula: Jour. Sed. Petrology, Vol. 55, p. 407-420.

Warren, J. K. (1988) Sedimentology of Coorong dolomite in the Salt Creek region, south Australia: Carbonates and Evaporites, Vol. 3, p. 175-200.

Warren, J. K. (1989) Evaporite Sedimentology: Prentice Hall, Englewood Cliffs, N.J., 285 pp.

Wiggins, W. D. and Harris, P. M. (1985) Burial diagenetic sequence in deep-water allochthonous dolomites, Permian Bone Spring Formation, southeast New Mexico: in Crevello, P. D. and Harris, P. M., eds., Deep-Water Carbonates, Soc. Economic Paleontologists and Mineralogists Core Workshop No. 6, p. 140-173.

Wiggins, W. D., Harris, P. M., and Burruss, R. C. (1993) Geochemistry of post-uplist calcite in the Permian Basin of Texas and New Mexico: Geol. Soc. Amer. Bull., Vol. 105, p. 779-790.

Wilson, A. O. (1985) Depositional and diagenetic facies in the Jurassic Arab-C and Arab-D reservoirs, Qatif Field, Saudi Arabia: in Roehl, P. O. and Choquette, P. W., eds., Carbonate Petroleum Reservoirs, Springer-Verlag, New York, p. 321-340.

Zenger, D. H. (1983) Burial dolomitization in the Lost Burro Formation (Devonian), east-central California, and the significance of late diagenetic dolomitization: Geology, Vol. 11, p. 519-522.

Zenger, D. H., and Dunham, J. B. (1988) Dolomitization of Siluro-Devonian limestones in a deep core (5,350 m), southeastern New Mexico: in Shukla, V. and Baker, P. A., eds., Sedimentology and Geochemistry of Dolostones, Soc. Econ. Paleontologists and Mineralogists Spec. Publ. 43, p. 161-173.

Zhang, C., Grossman, E. L., Yancey, T. E., and Mii, H.-S. (1990) Diagenesis and vital effect in brachiopod shells - the quest for the unaltered shell: Geol. Soc. Amer., Abs. with Prog., p. A116 (abs.).

APPENDICES

APPENDIX I. CONVENTIONS FOR REPORTING ISOTOPE DATA

Isotopes are species of a chemical element which differ from one another by atomic weight. Only stable, non-radioactive isotopes of oxygen, carbon and hydrogen are studied in dolomites and formation waters. Their abundances in nature are:

Oxygen	$^{18}O = 0.20\%$, $^{16}O = 99.76\%$;
Carbon	$^{13}C = 1.11\%$, $^{12}C = 98.89\%$
Hydrogen	$^{2}H = 0.02\%$, $^{1}H = 99.98\%$

These isotopes interest us because they vary in natural substances as a function of processes such as photosynthesis, evaporation, precipitation, cementation, dolomitization, and can therefore be used as tracers of these processes. The ratios of ^{18}O to ^{16}O in waters and rocks and of ^{13}C to ^{12}C in kerogen, oils and rocks are also of interest because they reveal diagenetic information. The ratio of ^{2}H or D (deuterium) to H is also of interest in water, hydrocarbon and hydrous mineral (e.g., clays) isotope studies.

The δ ("del") notation is used to express isotope proportions. $\delta^{18}O$, $\delta^{13}C$ and δD (sometimes expressed as $\delta^{2}H$) are defined as:

$$\delta^{18}O = [(^{18}O/^{16}O_{sample} - {}^{18}O/^{16}O_{standard}) / {}^{18}O/^{16}O_{standard}] * 1000 \quad \text{(i)}$$
$$\delta^{13}C = [(^{13}C/^{12}C_{sample} - {}^{13}C/^{12}C_{standard}) / {}^{13}C/^{12}C_{standard}] * 1000 \quad \text{(ii)}$$
$$\delta D = [(D/H_{sample} - D/H_{standard}) / D/H_{standard}] * 100 \quad \text{(iii)}$$

As shown by these equations, the δ value is the relative, not the absolute, difference in these ratios between a sample and a standard. Differences are expressed in parts per thousand for $\delta^{18}O$ and $\delta^{13}C$, or "per mil", symbolized by "‰", and in percent (%) for δD. If the $^{18}O/^{16}O$ ratio (or $^{13}C/^{12}C$, D/H) is smaller in the sample than in the standard, then the sample is ***depleted*** in ^{18}O (^{13}C, D) relative to the standard, and the δ value is a negative number. Conversely, if the $^{18}O/^{16}O$ ratio is larger in the sample than in the standard, then the sample is ***enriched*** in ^{18}O relative to the standard, and its δ value is a positive number. Substances which are depleted in the heavier isotope are referred to as "lighter" or just "depleted"; if they are enriched in the heavier isotopes, then they are "heavier" or "enriched".

The standards used to report the isotopic composition of water and carbonate minerals are SMOW (Standard Mean Ocean Water) and PDB (Peedee Formation *Belemnite*). The SMOW standard is commonly used to report the oxygen and deuterium isotopic compositions of fluids and of silicate minerals; the PDB standard is used to report the carbon and oxygen isotopic compositions of carbonate minerals. When comparing isotope data, it is best to report all data with respect to only one standard. This necessitates conversion of data in one standard notation to another (SMOW to PDB, or vice versa). To convert back and forth between the SMOW and PDB standards, use the following expressions:

$$\delta^{18}O(\text{SMOW Scale})=[(1.03086) *\delta^{18}O(\text{PDB Scale})]+30.86 \quad \text{(iv)}$$
$$\delta^{18}O(\text{PDB Scale})=[(0.97006) *\delta^{18}O(\text{SMOW Scale})]-29.94 \quad \text{(v)}$$

Consider a typical calcite oyster shell which is precipitated from seawater (SMOW) and has an $^{18}O/^{16}O$ ratio = .0020497 (SMOW itself has an $^{18}O/^{16}O$ equal to .0019997). Using Eq. (i), the calcite's $\delta^{18}O$ value is +25.0‰ on the SMOW scale:

$$[(.0020497-.0019997)/.0019997] * 1000 = +25.0‰$$

This illustrates that a calcite, with more ^{18}O than SMOW, has a positive $\delta^{18}O$ value on the SMOW scale. The oyster shell is isotopically heavier than seawater.

Using Eq. 2B, the oyster with a $\delta^{18}O$= +25‰ on the SMOW scale has a $\delta^{18}O$ of -5.7‰ on the PDB Scale, or less ^{18}O than the Cretaceous *Belemnite* standard:

$$\text{PDB value} = [(0.97006)*(25.0)]-29.94 = -5.7‰.$$

APPENDIX II. INTERPRETATION OF DOLOMITES BY COMPARISON TO MARINE CARBONATES

In order for a carbonate component to be used in constructing a marine isotopic baseline, one must be able to identify altered components and sample only unaltered components or the unaltered portions of partially altered components (Popp, et al., 1986a). Researchers working with the shells and skeletons of marine invertebrates and those working with marine carbonate cements have developed different approaches for obtaining original marine isotope values.

COMPARISON TO MARINE INVERTEBRATES

By only sampling the shells and skeletons of organisms that have retained 100% aragonite mineralogy or those which originally secreted shells of stable low-Mg calcite, many of the effects of diagenesis can be eliminated at the outset. Brachiopods, belemnites, and foraminifera are the taxa most commonly analyzed because of their original low-Mg calcite mineralogy. SEM studies designed to isolate samples with preserved microstructures also have proved useful (Brand and Morrison, 1987b). Trace element contents of fossil shells can be screened by comparing them to values for Recent equivalents. Those whose trace element contents have been altered by diagenesis can be eliminated prior to isotopic analysis (Brand and Morrison, 1987b). CL microscopy has been one of the most successful screening techniques. Marine cements and invertebrates contain little Mn and do not luminesce, whereas later diagenetic phases commonly contain Mn and luminesce. Nonluminescent areas are considered to best-represent unaltered marine material and luminescent areas can be avoided during microsampling, leading to acquisition of more pristine sample material for isotopic analysis (Popp, et al., 1986a). Application of these techniques allows a researcher to analyze "least-altered" samples and portions of samples. These are believed to yield the best estimates of original marine $\delta^{18}O$ and $\delta^{13}C$ composition. Because diagenesis in fresh water and at elevated temperature drive $\delta^{18}O$ negative, the most-enriched $\delta^{18}O$ values in a data set are commonly considered to be the "least-altered" when samples have not been screened by other techniques.

COMPARISON TO MARINE CARBONATE CEMENTS

Lohmann and coworkers have taken a different approach in their analysis of marine carbonate cements. Their technique is best-described in Given and Lohmann (1985). Under CL, former aragonite and high-Mg calcite cements are seen to be composed of intergrowths of luminescent and nonluminescent material. Each individual analysis of a cement sample represents a mixture with different proportions of nonluminescent (marine cement) and luminescent (later diagenetic) phases. Multiple analyses from a single sample define a linear covariant trend with a positive slope on a $\delta^{18}O$ vs $\delta^{13}C$ crossplot. $\delta^{18}O$ and $\delta^{13}C$ increase as the percentage of luminescent phase in the mixture decreases. When diagenetic trends for different localities are displayed on the same plot, the individual trends converge on a single $\delta^{18}O$ and $\delta^{13}C$ value, which is interpreted as the best estimate of the composition of nonluminescent calcite (i.e., the "least-altered" marine cement value; Fig. 15).

Lohmann and Walker (1989) have repeated this exercise for each period of the Phanerozoic. Their best estimate for the isotopic composition of marine carbonate for each geological period is presented as a single $\delta^{18}O$ and $\delta^{13}C$ value. Values for the entire Phanerozoic are summarized in Fig. 16. In Fig. 17 brachiopod data, collected by Popp, et al. (1986a) on samples screened for diagenetic alteration by CL, are compared to "least-altered" marine cement values, estimated by Lohmann and coworkers from the intersections of covariate populations of data. The two data sets agree in general, but there are several intervals in which the cement and invertebrate data diverge. Possible reasons for this are discussed in the following sections, and a more flexible approach to establishing marine carbonate isotopic baselines is presented.

LIMITATIONS OF THE TECHNIQUE

There are several problems inherent in representing the marine carbonate baseline for each geological period as a single isotopic value. A number of recent studies have demonstrated that a *range* of $\delta^{18}O$ values rather than a *single* $\delta^{18}O$ value is typical for each geological period, because variations in the temperature and salinity of seawater produce significant variations in the $\delta^{18}O$ of precipitated carbonate.

$\delta^{18}O$ of a marine fossil invertebrate is influenced by the latitude and depth of water at which it grew. The higher the latitude or deeper the water, the lower the ambient temperature, and the heavier the $\delta^{18}O$ of precipitated shell material (Brand, 1990; Grossman, et al., 1990). Salinity also has a profound effect on the $\delta^{18}O$ of shell material. Brachiopods can live in waters with salinities of 20 to 40‰. Because the $\delta^{18}O$ of water increases with salinity, a 20‰ range in salinity will result in significant variation in the $\delta^{18}O$ of brachiopods of equivalent age which lived in different environments (Brand, 1990). When varied together, temperature and salinity can have even more significant effects. Marine carbonates in epeiric seas may have precipitated from seawater that was hotter, more variable in salinity, and less stable thermally than open ocean water (Brand, 1990). $\delta^{18}O$ ranges for marine carbonates precipitated in epeiric seas would therefore be greater than those precipitated in the open ocean. Finally, the $\delta^{18}O$ of shell material precipitated from the same water at the same temperature may differ between species due to metabolic influence on fractionation (Zhang, et al., 1990). Each of these factors can lead to geographical variation in $\delta^{18}O$ among marine carbonates precipitated contemporaneously in different locations.

Superimposed on geographical variations in $\delta^{18}O$ are temporal variations that are known to occur within each geological period. Ludvigson, et al. (1990) identified a 3.5‰ increase in the $\delta^{18}O$ of brachiopod shells from Early to Late Ordovician times. Marshall and Middleton (1990) recognized a similar increase during the same time period and attributed it to a combination of "ice volume" effects coupled with decreasing temperature during the Late Ordovician glaciation. Adlis, et al. (1988) documented variations in $\delta^{18}O$ of >1‰ in brachiopods within individual glacio-eustatic cycles in Pennsylvanian shales. Lohmann and Walker (1989) reported an increase of 3‰ in the $\delta^{18}O$ of marine carbonate cements from the Late Devonian to the Early Mississippian, which they attributed to change in the $\delta^{18}O$ of seawater caused by change in the isotopic flux between silicate rocks and seawater. Thus, temporal variations in $\delta^{18}O$ *within* geological periods have been shown to be as great a 3.5‰.

CONSTRUCTION OF $\delta^{18}O$ BASELINES

Because of the geographical and temporal variations in $\delta^{18}O$ described above, we decided that a range of $\delta^{18}O$ values, rather than a single point, would best represent the marine carbonate baseline for each geological period. The baselines were constructed in this way: $\delta^{18}O$ (and $\delta^{13}C$) ranges for marine carbonate material reported in Table II-1 for brachiopods, belemnites, foraminifera, aragonite molluscs, aragonite ammonites, and "least-altered" marine carbonate cements, were plotted on $\delta^{18}O$ vs. $\delta^{13}C$ crossplots for each geological period. The total variation in $\delta^{18}O$ within each period represents the sum of all of the geological, biological, temporal, and diagenetic variations summarized above. The heaviest $\delta^{18}O$ value for each individual data range was selected as the "least-altered" value for that particular data set.[1] The most positive $\delta^{18}O$ value in the heaviest data range and the most positive $\delta^{18}O$ value in the lightest data range were chosen as the upper and lower limits of the marine carbonate isotopic baseline for each period. These most-enriched values provide an estimate of the range of "least-altered" marine carbonate material within each geological period and are displayed as dashed lines on Figs. II-1 through II-14. These values were used to establish the shaded marine baselines shown on Figs. 19A, 19B, and 22A. In several periods (the Ordovician, Devonian, and Cretaceous) unusually heavy $\delta^{18}O$ values were reported from single localities. These ^{18}O-enriched data sets appeared anomalous compared to all other data sets from each of these periods. In each case, the upper limit for $\delta^{18}O$ was readjusted to a value more consistent with the heavy end of the rest of data for that period (Figs. II-4, II-11, and II-13).

Although the $\delta^{18}O$ ranges of marine carbonate baselines might be narrowed by construction of marine baseline plots at the epoch rather than the period level, sufficient data are unavailable for doing so for many epochs in the Phanerozoic at the present time.

CONSTRUCTION OF $\delta^{13}C$ BASELINES

The $\delta^{13}C$ compositions of early dolomite varies widely depending on the relative proportions of carbon derived from oxidized organic material (which is isotopically light), precursor limestone (which is isotopically intermediate), and carbon from methanogenic reactions (which is isotopically heavy) that are incorporated into the carbonate lattice during dolomitization (Fig. 14). Because of this wide variability in the $\delta^{13}C$ compositions of early dolomite, $\delta^{13}C$ is less useful than $\delta^{18}O$ for differentiating low temperature and high temperature dolomites. Only $\delta^{18}O$ baselines have been constructed in this Appendix, but the same technique described above for generating $\delta^{18}O$ baselines can be used to generate $\delta^{13}C$ baselines for each geological period using the carbon isotope ranges listed in Table II-1.

[1]This, of course, is an oversimplification, because the more negative $\delta^{18}C$ values within each population may be caused not only by diagenetic alteration but also by natural variation in $\delta^{18}O$ within the population. However, it is the only practical way to eliminate the effects of diagenetic alteration when comparing data collected on so many different types of samples.

Table II-1. Carbon and oxygen isotope ranges of marine carbonate material for each geological period.

Age	Reference	Description	Location	$\delta^{18}O$ ‰ PDB	$\delta^{13}C$ ‰ PDB
Quaternary (Holocene)	Gonzales & Lohmann (1985)	Abiotic marine aragonite & HMC cements	Bermuda	+0.7 to +1.3	+3.7 to +4.9
Quaternary (Holocene)	Gonzales & Lohmann (1985)	Molluscs & foraminifera	Bermuda	-0.4 to +0.7	+0.1 to +3.5
Quaternary (Holocene)	Gonzales & Lohmann (1985)	Abiotic marine aragonite & HMC cements	Enewetak	-1.0 to -0.5	+4.0 to +4.2
Quaternary (Holocene)	Gonzales & Lohmann (1985)	Molluscs & foraminifera	Enewetak	-2.3 to -1.5	-2.1 to +3.5
Quaternary (Holocene)	Gonzales & Lohmann (1985)	Abiotic marine aragonite & HMC cements	Bikini	-0.7 to -0.3	+3.2 to +4.5
Quaternary (Holocene)	Gonzales & Lohmann (1985)	Molluscs & foraminifera	Bikini	-2.8 to -2.2	-1.1 to +1.3
Quaternary (Holocene)	Gonzales & Lohmann (1985)	Abiotic marine aragonite & HMC cements	Belize	+1.2 to +2.0	+2.8 to +4.2
Quaternary (Holocene)	Gonzales & Lohmann (1985)	Molluscs & foraminifera	Belize	+0.2 to +2.2	+2.0 to +4.1
Quaternary (Holocene)	Kaufman & Magaritz (1980)	Whole pelecypods	Mediterranean coast, Israel	-2.0 to +1.0	NA
Quaternary (Holocene)	Lowenstam (1961)	Whole brachiopods which showed good preservation of original microstructures	Atlantic & Pacific Oceans & Mediterranean Sea	-1.9 to +2.0	NA
Quaternary (Holocene)	Lohmann & Walker (1989)	Most-enriched isotopic values for abiotic marine calcite cements	NA	-0.8	+3.0
Quaternary (Holocene)	James & Ginsburg (1979)	Botryoidal aragonite	Belize reef	+0.4 to +1.1	+3.3 to +4.3
Quaternary (Holocene)	James & Choquette (1983)*	Heaviest marine cement compositions from covariate populations of data	Belize	+1.0	+3.8
Quaternary (Pleistocene)	Brand & Morrison (1987b)*	Whole molluscs (mostly aragonite)	NA	-11.8 to +1.7	-5.5 to +2.4
Quaternary (Pleistocene)	Savin, et al. (1975)	Open-ocean planktonic foraminifera	Low latitude DSDP sites, N Pacific Ocean	-0.8 to +1.2	+1.1 to +1.8
Quaternary (Pleistocene)	Lloyd (1969)	Aragonite pelecypods	Caloosahatchee Formation, FL, USA	-1.9 to +3.6	-2.9 to +3.6
Tertiary (Pliocene)	Lowenstam (1961)	Whole brachiopods which showed good preservation of original microstructures	Careaga Ss., CA, USA	+0.7	NA
Tertiary (Pliocene)	Brand & Morrison (1987b)*	Whole molluscs (mostly aragonite)	NA	-0.6 to 0.0	-0.5 to +0.1

Table II-1. Carbon and oxygen isotope ranges of marine carbonate material for each geological period.

Age	Reference	Description	Location	$\delta^{18}O$ ‰ PDB	$\delta^{13}C$ ‰ PDB
Tertiary (Pliocene)	Buchardt (1978)	Whole molluscs (mostly aragonite)	NW Europe	-0.8 to +1.9	NA
Tertiary (Pliocene)	Savin, et al. (1975)	Open-ocean planktonic foraminifera	Low latitude DSDP sites, N Pacific Ocean	-0.8 to +1.1	+0.9 to +2.0
Tertiary (Miocene)	Brand & Morrison (1987b)*	Whole molluscs (mostly aragonite)	NA	-1.4 to +2.6	-3.7 to +2.4
Tertiary (Miocene)	Buchardt (1978)	Whole molluscs (mostly aragonite)	NW Europe	-1.5 to +3.1	NA
Tertiary (Miocene)	Savin, et al. (1975)	Open-ocean planktonic foraminifera	Low latitude DSDP sites, N Pacific Ocean	-1.1 to +1.1	+1.6 to +2.9
Tertiary (Miocene)	Miller, et al. (1987)	Tropical & subtropical surface-dwelling planktonic foraminifera	DSDP sites, Atlantic & Indian Oceans	-1.3 to +0.3	NA
Tertiary (Oligocene)	Brand & Morrison (1987b)*	Whole molluscs (mostly aragonite)	NA	-3.1 to +1.5	-5.2 to +3.5
Tertiary (Oligocene)	Buchardt (1978)	Whole molluscs (mostly aragonite)	NW Europe	-1.9 to +3.0	NA
Tertiary (Oligocene)	Miller, et al. (1987)	Tropical & subtropical surface-dwelling planktonic foraminifera	DSDP sites, Atlantic & Indian Oceans	+0.1 to +1.8	NA
Tertiary (Oligocene)	Savin, et al. (1975)	Open-ocean planktonic foraminifera	Low latitude DSDP sites, N Pacific Ocean	-0.6 to +0.4	+0.7 to +2.0
Tertiary (Eocene)	Brand & Morrison (1987b)*	Whole molluscs (mostly aragonite)	NA	-2.4 to -1.8	+0.8 to +1.4
Tertiary (Eocene)	Buchardt (1978)	Whole molluscs (mostly aragonite)	NW Europe	-4.7 to -1.5	NA
Tertiary (Eocene)	Savin, et al. (1975)	Open-ocean planktonic foraminifera	Low latitude DSDP sites, N Pacific Ocean	-1.7 to +0.2	+1.1 to +2.3
Tertiary (Paleocene)	Brand & Morrison (1987b)*	Whole molluscs (mostly aragonite)	NA	-1.8 to +1.3	-3.1 to +1.9
Tertiary (Paleocene)	Buchardt (1978)	Whole molluscs (mostly aragonite)	NW Europe	-3.2 to +0.5	NA
Tertiary (Paleocene)	Savin, et al. (1975)	Open-ocean planktonic foraminifera	Low latitude DSDP sites, N Pacific Ocean	-1.7 to -1.0	+2.2 to +3.9
Upper Cretaceous	Lowenstam (1961)	Whole brachiopods which showed good preservation of original microstructures	Navesink Marl, NJ, USA	-0.6 to -0.4	NA

Table II-1. Carbon and oxygen isotope ranges of marine carbonate material for each geological period.

Age	Reference	Description	Location	$\delta^{18}O$ ‰ PDB	$\delta^{13}C$ ‰ PDB
Upper Cretaceous	Al-Aasm & Veizer (1986)	Aragonite caprinid rudists	Coon Creek Tongue-Ripley Fm., TX, USA	-2.3 to -2.1	+3.2 to +3.9
Upper Cretaceous (Maastrichtian)	Savin, et al. (1975)	Open-ocean planktonic foraminifera	Low latitude DSDP sites, N Pacific Ocean	-1.4 to -0.8	+1.2 to +2.5
Upper Cretaceous (Maastrichtian)	Buchardt & Weiner (1981)	Well preserved ammonites (>98% aragonite)	Greenland	-2.5 to +0.4	-3.7 to +2.2
Upper Cretaceous (Maastrichtian)	Lowenstam & Epstein (1954)	Belemnite guards showing no evidence of recrystallization or contamination by secondary calcite	W Europe	-0.7 to +0.8	NA
Upper Cretaceous (Campanian to Maastrichtian)	Morrison & Brand (1988)	Well-preserved aragonite molluscs	Lea Park Fm., W Canada	-7.9 to +0.9	-6.6 to +5.5
Upper Cretaceous (Campanian to Maastrichtian)	Morrison & Veizer (1990)	Preserved belemnites	Holland & Antarctica	-0.2 to +0.6	NA
Upper Cretaceous (Campanian)	Lowenstam & Epstein (1954)	Belemnite guards showing no evidence of recrystallization or contamination by secondary calcite	W Europe	-1.0 to +0.2	NA
Upper Cretaceous (Campanian)	Buchardt & Weiner (1981)	Well preserved ammonites (>98% aragonite)	Greenland	-3.7 to -1.0	-1.9 to +0.4
Upper Cretaceous (Lower Campanian)	Lowenstam (1961)	Whole brachiopods which showed good preservation of original microstructures	Ivoe, Sweden	-1.2 to -0.8	NA
Upper Cretaceous (Conician & Santonian)	Czerniakowski, et al. (1984)	Non-luminescent portions of LMC oyster shells	Austin Chalk, TX	-4.7 to -1.5	-0.5 to +3.2
Upper Cretaceous (Conician & Santonian)	Lowenstam & Epstein (1954)	Belemnite guards showing no evidence of recrystallization or contamination by secondary calcite	W Europe	-1.4 to -0.4	NA
Upper Cretaceous (Turonian)	Lowenstam & Epstein (1954)	Belemnite guards showing no evidence of recrystallization or contamination by secondary calcite	W Europe	-0.7 to 0.0	NA

Table II-1. Carbon and oxygen isotope ranges of marine carbonate material for each geological period.

Age	Reference	Description	Location	$\delta^{18}O$ ‰ PDB	$\delta^{13}C$ ‰ PDB
Upper Cretaceous (Cenomanian)	Lowenstam & Epstein (1954)	Belemnite guards showing no evidence of recrystallization or contamination by secondary calcite	W Europe	-0.1 to +0.2	NA
Cretaceous	Brand & Morrison (1987b)*	Whole brachiopods & belemnite guards	NA	-5.3 to +0.7	-3.2 to +3.2
Cretaceous	Brand & Morrison (1987b)*	Whole molluscs (mostly aragonite)	NA	-4.6 to +0.5	-4.0 to +3.4
Middle Cretaceous	James & Choquette (1983)*	Heaviest marine cement compositions from covariate populations of data	NE Mexico	-2.3	+2.0
Lower Cretaceous (Albian)	Lowenstam & Epstein (1954)	Belemnite guards showing no evidence of recrystallization or contamination by secondary calcite	Worldwide	-1.7 to +0.2	NA
Lower Cretaceous (Valanginian to Barremian)	Morrison & Veizer (1990)	Preserved belemnites	Germany	-2.1 to +3.5	NA
Lower Cretaceous	Moldovanyi & Lohmann (1984)	Rudists & micrites	Sligo & Cupido Fms., S TX & NE Mexico	-5.0 to -2.0	+1.8 to +4.0
Lower Cretaceous	Moldovanyi & Lohmann (1984)	Isopachous marine cements	Sligo & Cupido Fms., S TX & NE Mexico	-4.5 to -1.8	+2.0 to +3.5
Lower Cretaceous	Lohmann & Walker (1989)	Most-enriched isotopic values for abiotic marine calcite cements	NA	-2.0	+3.0
Lower Cretaceous	James & Choquette (1983)*	Heaviest marine cement compositions from covariate populations of data	S Texas, USA	-2.0	+2.5
Upper Jurassic	Lohmann & Walker (1989)	Most-enriched isotopic values for abiotic marine calcite cements	NA	-1.3	+3.0
Upper Jurassic (Oxfordian)	Veizer & Fritz (1976)	Least-altered belemnite rostra (<5% altered)	S Germany	-1.0	+0.6
Jurassic	Brand & Morrison (1987b)*	Belemnite guards	NA	-3.8 to +0.7	-1.8 to +2.0
Jurassic	Brand & Morrison (1987b)*	Whole molluscs (mostly aragonite)	NA	-6.5 to +0.2	-1.8 to +4.0
Jurassic	James & Choquette (1983)*	Heaviest marine cement compositions from covariate populations of data	England	-0.2	+2.8

Table II-1. Carbon and oxygen isotope ranges of marine carbonate material for each geological period.

Age	Reference	Description	Location	$\delta^{18}O$ ‰ PDB	$\delta^{13}C$ ‰ PDB
Jurassic	Tan & Hudson (1974)	Preserved aragonite and calcite shells of benthonic bivalves and brachiopods from brackish to open marine environments (freshwater species excluded)	Great Estuarine Series, Scotland	-6.2 to -0.1	-2.6 to +3.8
Middle Jurassic (Callovian)	Brand (1986)	Aragonite ammonoids	Lukow Sh., Poland	0 to +0.8	-1.6 to +2.1
Middle Jurassic (Callovian)	Tan, et al. (1970)	Well-preserved belemnites	Belemnite Ss. & Oxford Clay, Scotland	-1.9 to +0.5	+0.9 to +2.7
Middle Jurassic (Callovian)	Tan, et al. (1970)	Aragonite ammonites	Oxford Clay, Scotland	-2.9 to -2.6	+2.6 to +3.1
Middle Jurassic	Marshall & Ashton (1980)	Hardground cemented by acicular marine cements	Lincolnshire Ls., England	-3.6 to -0.2	+0.3 to +3.1
Middle Jurassic (Dogger)	Veizer & Fritz (1976)	Least-altered belemnite rostra (<5% altered)	S Germany	-3.1 to +0.6	-0.3 to +1.6
Lower Jurassic (Lias)	Veizer & Fritz (1976)	Least-altered belemnite rostra (<5% altered)	S Germany	-2.7 to -0.4	0 to +1.3
Upper Triassic	Lohmann & Walker (1989)	Most-enriched isotopic values for abiotic marine calcite cements	NA	-0.5	+2.6
Triassic	James & Choquette (1983)*	Heaviest marine cement compositions from covariate populations of data	Sicily	-0.8	+3.0
Late Permian	Popp, et al. (1986a)*	Estimates of least-altered marine cements (from Lohmann & coworkers)	NA	-3.3 to -2.3	+4.5 to +5.5
Late Permian	Popp, et al. (1986a)	Non-luminescent portions of brachiopods	NA	+0.8 to +1.8	+3.0 to +4.0
Late Permian (Late Guadalupian)	Given & Lohmann (1985)	Former aragonite marine cements	Capitan Fm., TX, USA	-9.7 to -1.9	-1.2 to +5.4
Upper Permian	Gruszczynski, et al. (1989)	Productacea brachiopods with fully preserved primary microstructures	Kapp Starostin Fm., Spitsbergen	-13.5 to -2.8	-4.9 to +7.3
Upper Permian	James & Choquette (1983)*	Heaviest marine cement compositions from covariate populations of data	W Texas, USA	-1.8	+5.5
Upper Permian	Lohmann & Walker (1989)	Most-enriched isotopic values for abiotic marine calcite cements	NA	-2.6	+4.2
Permian	Brand & Morrison (1987b)*	Whole brachiopods	NA	-10.5 to -1.0	-3.6 to +5.8

Table II-1. Carbon and oxygen isotope ranges of marine carbonate material for each geological period.

Age	Reference	Description	Location	$\delta^{18}O$ ‰ PDB	$\delta^{13}C$ ‰ PDB
Permian	James & Choquette (1983)*	Heaviest marine cement compositions from covariate populations of data	NE Canada	-2.0	+5.2
Permian	Veizer, et al. (1986)	Whole brachiopods	W Canada	-10.6 to -1.0	-3.6 to +5.2
Lower Permian	Lowenstam (1961)	Whole brachiopods which showed good preservation of original microstructures	Noonkanbah & Lyons Fms., Australia	-3.7 to +2.2	NA
Early Permian	Popp, et al. (1986a)*	Estimates of least-altered marine cements (from Lohmann & coworkers)	NA	-3.5 to -1.5	+4.5 to +6.5
Early Permian	Popp, et al. (1986a)	Whole brachiopods	NA	-7.3 to +2.0	-2.6 to +6.2
Early Permian	Popp, et al. (1986a)	Non-luminescent portions of brachiopods	NA	-6.5 to +0.3	+2.7 to +6.8
Pennsylvanian to Lower Permian	Davies & Krouse (1975)	Marine carbonate cements, brachiopods, and marine internal sediments	Nansen & Hare Fiord Fms., Canada	-8.0 to -1.0	+4.0 to +7.2
Late Carboniferous	Popp, et al. (1986a)	Whole brachiopods	NA	-9.8 to +0.5	-5.7 to +4.2
Late Carboniferous	Popp, et al. (1986a)	Non-luminescent portions of brachiopods	NA	-4.8 to -1.1	+4.5 to +6.2
Late Carboniferous	Popp, et al. (1986a)*	Estimates of least-altered marine cements (from Lohmann & coworkers)	NA	-3.5 to -1.5	+4.5 to +6.5
Upper Pennsylvanian (Virgilian)	Adlis, et al. (1988)	Well-preserved, nonluminescent fibrous calcite in brachiopod shells	Necessity & Colony Creek Sh., N TX, USA	-4.2 to -0.1	-1.1 to +3.8
Upper Pennsylvanian	Brand & Morrison (1987a)	Brachiopods & aragonite cephalopods	Brush Ck. Mbr., Glenshaw Fm., PA, USA	-7.7 to -4.4	-2.1 to +5.9
Pennsylvanian	James & Choquette (1983)*	Heaviest marine cement compositions from covariate populations of data	New Mexico, USA	-2.0	+4.6
Middle Pennsylvanian (Desmoinesian)	Brand (1989a)	Aragonite cephalopods & bivalves	Boggy Fm. (Buckhorn Alphalt), OK, USA	-4.2 to +0.5	0.0 to +7.0
Middle Pennsylvanian	Brand (1987)	Least-altered brachiopods & molluscs	Magoffin Sh., KY, USA	-5.6 to -4.2	NA
Middle Pennsylvanian	Lohmann & Walker (1989)	Most-enriched isotopic values for abiotic marine calcite cements	NA	-2.0	+3.4

Table II-1. Carbon and oxygen isotope ranges of marine carbonate material for each geological period.

Age	Reference	Description	Location	$\delta^{18}O$ ‰ PDB	$\delta^{13}C$ ‰ PDB
Lower Pennsylvanian	Brand (1981)	Unaltered calcite & aragonite brachiopods & molluscs	Kendrick Sh., KY, USA	-5.1 to -4.0	-5.4 to +2.3
Carboniferous	Brand & Morrison (1987b)*	Whole molluscs (mostly aragonite)	NA	-7.2 to +0.5	-4.5 to +7.0
Carboniferous	Brand & Morrison (1987b)*	Whole brachiopods	NA	-8.7 to -2.3	-3.2 to +6.0
Carboniferous	Brand (1989b)	Least-altered brachiopods from SEM & trace element analysis	North America	-7.0 to -0.6	-3.3 to +4.6
Carboniferous	Veizer, et al. (1986)	Whole brachiopods	W Canada	-8.6 to -3.1	-2.6 to +5.7
Upper Mississippian	Lowenstam (1961)	Whole brachiopods which showed good preservation of original microstructure	Goddard Sh., OK, USA	-3.0 to -1.8	NA
Mississippian	Brand & Veizer (1981)	Least-altered brachiopods & rugose corals	Burlington Ls., IO & MO, USA	-5.2 to -2.3	+0.1 to +4.3
Mississippian	James & Choquette (1983)*	Heaviest marine cement compositions from covariate populations of data	New Mexico, USA	-1.2	+3.8
Early Carboniferous	Dunn, (1988)	Most-enriched isotopic values for abiotic marine calcite cements	Waulsortian reefs, SW Ireland	-11.6 to -0.7	+1.9 to +4.3
Early Carboniferous	Popp, et al. (1986a)*	Estimates of least-altered marine cements (from Lohmann & coworkers)	NA	-2.0 to -1.0	+3.5 to +4.5
Early Carboniferous	Popp, et al. (1986a)	Whole brachiopods	NA	-12.0 to -1.6	-4.1 to +4.3
Early Carboniferous	Popp, et al. (1986a)	Non-luminescent portions of brachiopods	NA	-11.3 to -0.8	-5.2 to +3.8
Lower Mississippian	Lohmann & Walker (1989)	Most-enriched isotopic values for abiotic marine calcite cements	NA	-1.1	+4.0
Early Mississippian (Osagean)	Meyers and Lohmann (1985)	Crinoids, former HMC marine cements, and marine micrites.	Lake Valley & Kelly Fms., NM, USA	-6.7 to -1.4	+0.6 to +4.8
Early Mississippian (Osagean)	Brand (1982)	Whole brachiopods	Lake Valley Fm., NM, USA	-2.3 to -1.7	+3.0 to +3.5
Early Mississippian (Osagean)	Brand (1982)	Whole brachiopods	Burlington Fm., IO & MO, USA	-4.8 to -2.3	+2.8 to +4.2
Upper Devonian	James & Choquette (1983)*	Heaviest marine cement compositions from covariate populations of data	Alberta, Canada	-4.5 & -4.4	+2.5 & +2.0

Table II-1. Carbon and oxygen isotope ranges of marine carbonate material for each geological period.

Age	Reference	Description	Location	$\delta^{18}O$ ‰ PDB	$\delta^{13}C$ ‰ PDB
Upper Devonian (Famennian)	Hurley & Lohmann (1989)	Marine carbonate cements & Neptunian dikes	Canning Basin, Australia	-7.5 to -4.6	-0.5 to +2.8
Upper Devonian (Frasnian)	Carpenter & Lohmann (1989)	Clear radiaxial fibrous marine cements	Golden Spike & Nevis reefs, Alberta, Canada	-6.0 to -4.3	+0.6 to +3.4
Upper Devonian (Frasnian)	Hurley & Lohmann (1989)	Marine carbonate cements, brachiopods, & Neptunian dikes	Canning Basin, Australia	-7.2 to -4.0	+1.6 to +3.0
Upper Devonian (Frasnian)	Popp, et al. (1986a)	Whole brachiopods	NA	-6.2 to -3.7	-0.2 to +1.9
Late Devonian (Frasnian)	Dunn, (1988)	Most-enriched isotopic values for abiotic marine calcite cements	Dinant Synclinorium, Belgium	-11.9 to -3.9	0.0 to +5.5
Late Devonian	Popp, et al. (1986a)*	Estimates of least-altered marine cements (from Lohmann & coworkers)	NA	-6.5 to -5.5	+1.3 to +2.3
Upper Devonian	Veizer, et al. (1986)	Whole brachiopods	NWT, Canada	-6.4 to -5.6	+0.4 to +1.4
Upper Devonian	Mattes & Mountjoy (1980)	Radiaxial fibrous marine calcite cement and marine internal sediment	Miette Buildup, Alberta, Canada	-13.3 to -5.1	-2.0 to +2.7
Upper Devonian	Mountjoy and Walls (1977)	Radiaxial fibrous marine calcite cement and marine internal sediment	Golden Spike Reef, Alberta, Canada	-7.7 to -5.8	+1.6 to +3.5
Upper Devonian	Lohmann & Walker (1989)	Most-enriched isotopic values for abiotic marine calcite cements	NA	-5.0	+2.3
Devonian	Brand & Morrison (1987b)*	Whole brachiopods	NA	-10.7 to -4.4	-1.8 to +4.4
Middle Devonian	Veizer, et al. (1986)	Whole brachiopods	NWT, Canada	-8.9 to -4.6	-1.9 to +2.7
Middle Devonian	Lohmann & Walker (1989)	Most-enriched isotopic values for abiotic marine calcite cements	NA	-4.5	+1.4
Middle Devonian	Popp, et al. (1986a)*	Estimates of least-altered marine cements (from Lohmann & coworkers)	NA	-5.0 to -4.0	+1.0 to +2.0
Middle Devonian (Givetian)	Popp, et al. (1986a & b)	Whole brachiopods & nonluminescent portions of brachiopods	Hamilton Grp., NE USA & S Canada	-12.5 to -2.5	-4.5 to +4.9
Lower/Middle Devonian	Veizer, et al. (1986)	Whole brachiopods	NWT, Canada	-8.3 to -4.4	-0.8 to +1.6
Lower Devonian	Veizer, et al. (1986)	Whole brachiopods	NWT, Canada	-13.4 to -4.1	-1.2 to +0.4

Table II-1. Carbon and oxygen isotope ranges of marine carbonate material for each geological period.

Age	Reference	Description	Location	$\delta^{18}O$ ‰ PDB	$\delta^{13}C$ ‰ PDB
Lower Devonian	Lohmann & Walker (1989)	Most-enriched isotopic values for abiotic marine calcite cements	NA	-3.8	+1.4
Late Silurian	Popp, et al. (1986a)	Whole brachiopods	NA	-7.2 to -3.2	-0.5 to +3.0
Upper Silurian	Lohmann (1988a)	Most-enriched isotopic values for abiotic marine calcite cements	Pipe Creek Jr. reef, IN, USA	-4.0 to -3.6	+1.6 to +2.0
Upper Silurian	Lohmann & Walker (1989)	Most-enriched isotopic values for abiotic marine calcite cements	NA	-4.0	+1.4
Silurian	Brand & Veizer (1981)	Least-altered brachiopods & rugose corals	Arctic Islands, Canada	-6.7 to -5.0	-0.3 to +1.4
Silurian	Brand & Morrison (1987b)*	Whole brachiopods	NA	-7.8 to -4.4	-3.0 to +3.8
Silurian	Veizer, et al. (1986)	Whole brachiopods	Quebec & NWT, Canada	-7.8 to -4.4	-3.1 to +3.8
Silurian	James & Choquette (1983)*	Heaviest marine cement compositions from covariate populations of data	Michigan, USA	-5.4 & -4.0	+1.2 & +1.3
Middle Silurian (Niagaran)	Cercone & Lohmann (1985)	Former Mg-calcite marine carbonate cements	Niagaran reefs, Michigan, USA	-8.2 to -3.5	+1.2 to +5.4
Late Ordovician	Marshall & Middleton (1990)	Radiaxial marine carbonate cements & Non-luminescent portions of brachiopods	Kullsberg & Boda Ls., Sweden	-6.6 to -1.5	0 to +6.9
Late Ordovician	Popp, et al. (1986a)*	Estimates of least-altered marine cements (from Lohmann & coworkers)	NA	-5.0 to -4.0	-1.0 to 0.0
Late Ordovician (Ashgillian)	Popp, et al. (1986a)	Non-luminescent portions of brachiopods	NA	-5.1 to -4.0	+0.1 to +1.4
Late Ordovician (Caradocian)	Popp, et al. (1986a)	Whole brachiopods	NA	-6.7 to -5.4	+0.1 to +1.8
Upper Ordovician	Lohmann & Walker (1989)	Most-enriched isotopic values for abiotic marine calcite cements	NA	-4.5	0.0
Ordovician	Brand & Morrison (1987b)*	Whole brachiopods	NA	-5.5 to -4.5	+0.7 to +1.3
Ordovician	James & Choquette (1983)*	Heaviest marine cement compositions from covariate populations of data	Nevada, USA	-6.6	+0.3
Ordovician	Veizer, et al. (1986)	Whole brachiopods	Quebec, Canada	-5.5 to -4.5	+0.5 to +1.2

Table II-1. Carbon and oxygen isotope ranges of marine carbonate material for each geological period.

Age	Reference	Description	Location	$\delta^{18}O$ ‰ PDB	$\delta^{13}C$ ‰ PDB
Middle Ordovician	Popp, et al. (1986a)*	Estimates of least-altered marine cements (from Lohmann & coworkers)	NA	-7.0 to -6.0	0.0 to +1.0
Middle Ordovician	Lohmann & Walker (1989)	Most-enriched isotopic values for abiotic marine calcite cements	NA	-6.6	+0.4
Lower Ordovician	Lohmann & Walker (1989)	Most-enriched isotopic values for abiotic marine calcite cements	NA	-5.0	-1.2
Early Ordovician	Popp, et al. (1986a)*	Estimates of least-altered marine cements (from Lohmann & coworkers)	NA	-6.5 to -5.5	-1.5 to -0.5
Middle Cambrian	Lohmann & Walker (1989)	Most-enriched isotopic values for abiotic marine calcite cements	NA	-5.0	-0.7
Lower Cambrian	James & Klappa (1983)	Synsedimentary fibrous calcite cement	Forteau Fm., Labrador, Canada	-6.2 to -4.8	-0.4 to -0.2
Lower Cambrian	James & Choquette (1983)*	Heaviest marine cement compositions from covariate populations of data	Labrador, Canada	-4.7	-0.1

*These papers interpret data originally published in other papers shown in this table.

**This paper contains brachiopod, belemnite, and mollusc data for most periods of the Phanerozoic which were assembled from other papers, many of which appear in this table.

Table II-2. Carbon and oxygen isotope ranges for Quaternary early low temperature dolomites shown in Figure 18.

Reference	Location	Age of Host Sediment	Age of Dolomitization	Environment of Dolomitization	$\delta^{18}O$ ‰ PDB	$\delta^{13}C$ ‰ PDB
(a) Kinsman and Patterson (1973)	Abu Dhabi	Holocene	Holocene	Sabkha	+2.5 to +3.7	+4.0 to +4.4
(b) McKenzie (1981)	Abu Dhabi	Holocene	Holocene	Sabkha	-0.4 to +3.9	+2.0 to +3.7
(c) Mazzullo, et al. (1987)	Belize	Holocene	Holocene	Supratidal	+1.2 to +2.2	-5.2 to -0.6
(d) Carballo, et al. (1987)	Sugarloaf Key, Florida, USA	Holocene	Holocene	Supratidal	+2.4 to +3.2	-2.4 to +1.1
(e) Clayton, et al. (1968)	Coorong Region, Australia	Holocene	Holocene	Coastal lagoons and ephemeral lakes	+3.9 to +4.2	+0.1 to +4.5
(f) Rosen, et al. (1988)	Coorong Region, Australia	Holocene	Holocene	Coastal ephemeral lakes	+4.6 to +8.1	-1.1 to +4.3
(g) Mitchell, et al. (1987)	Jamaica	Holocene	Holocene	Subtidal marine	+2.0	+1.6
(h) Behrens & Land (1972)	Baffin Bay, Texas, USA	Holocene	Holocene (^{14}C dated)	Subtidal, elevated salinities.	+4.1 to +5.2	-4.1 to +4.2
(i) Kier (1973)	Baffin Bay, Texas, USA	Holocene	Holocene (^{14}C dated)	Subtidal, possibly hypersaline	+4.8 to +5.4	-7.5 to -1.6
(j) Magaritz, et al. (1980)	Coastal Plain, Israel	Holocene	Pleistocene to Holocene (^{14}C dated)	Marine – meteoric mixing zone	-1.3	-9.6
(k) Land (1973a)	Falmouth Fm., Jamaica	Holocene	Late Pleistocene (whole sed. ^{14}C dated)	Marine – meteoric mixing zone	-1.6 to -0.9	-8.7 to -7.1
(l) Land (1973b)	Hope Gate Fm., Jamaica	Pleistocene	Pleistocene	Inferred marine – meteoric mixing zone	+0.4 to +3.2	-1.8 to +3.8
(m) Humphrey (1988)	Barbados	Pleistocene	Pleistocene	Marine – meteoric mixing zone	+1.0 to +3.6	-23.5 to -9.5
(n) Ward & Hailey (1985)	Yucatan	Pleistocene	Pleistocene	Marine – meteoric mixing zone	+0.5 to +3.2	+1.1 to +2.4
(o) Supko (1977) & Swart, et al. (1987)	San Salvador, Bahamas	Pliocene (only)	Pleistocene	Normal marine or marine – meteoric mixing zone	+0.4 to +4.1	+0.3 to +2.6

Holocene

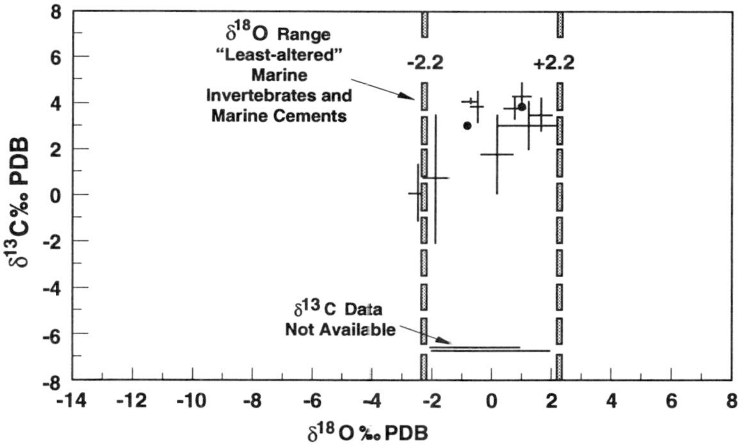

Figure II-1

Pleistocene

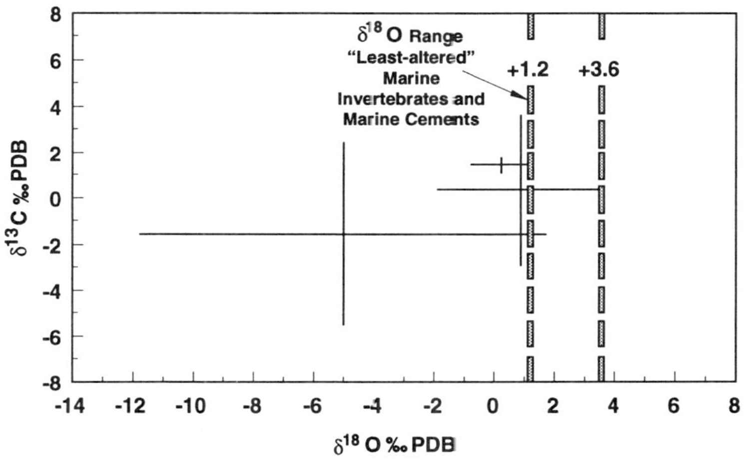

Figure II-2

Figures II-1. to II-14. δ^{18}O values for marine invertebrates and marine carbonate cements are shown for each period of the Phanerozoic and for the Pleistocene and Holocene epochs of the Quaternary Period. The data from which these figures were constructed are compiled in Table II-2. δ^{18}O and δ^{13}C ranges from a single paper or locality are shown as intersecting horizontal and vertical bars. Many papers report only δ^{18}O ranges. These are plotted at the bottoms of figures as horizontal bars. Their positions with respect to the δ^{13}C axis have no significance. Other papers report only "least-altered" δ^{18}O and δ^{13}C values. These are shown as points.

Since δ^{18}O of marine carbonate material varies within each geological period due to water temperature and salinity variations, fluctuation in δ^{18}O of seawater caused by glaciation, and metabolic effects in organisms, choosing a single δ^{18}O value to represent each period is unrealistic. "Least-altered" δ^{18}O values are therefore shown as ranges rather than as points. The vertical, dashed lines represent reasonable limits for the δ^{18}O values expected for primary marine carbonates of a particular geological age which have not been altered by diagenesis. The distance between the dashed lines represents the best estimate of the δ^{18}O range of unaltered marine carbonate material for that period.

Three papers report anomalously heavy data from single localities. These are plotted but were not used in determining the upper limits of δ^{18}O (Figs. II-4, II-11, and II-13).

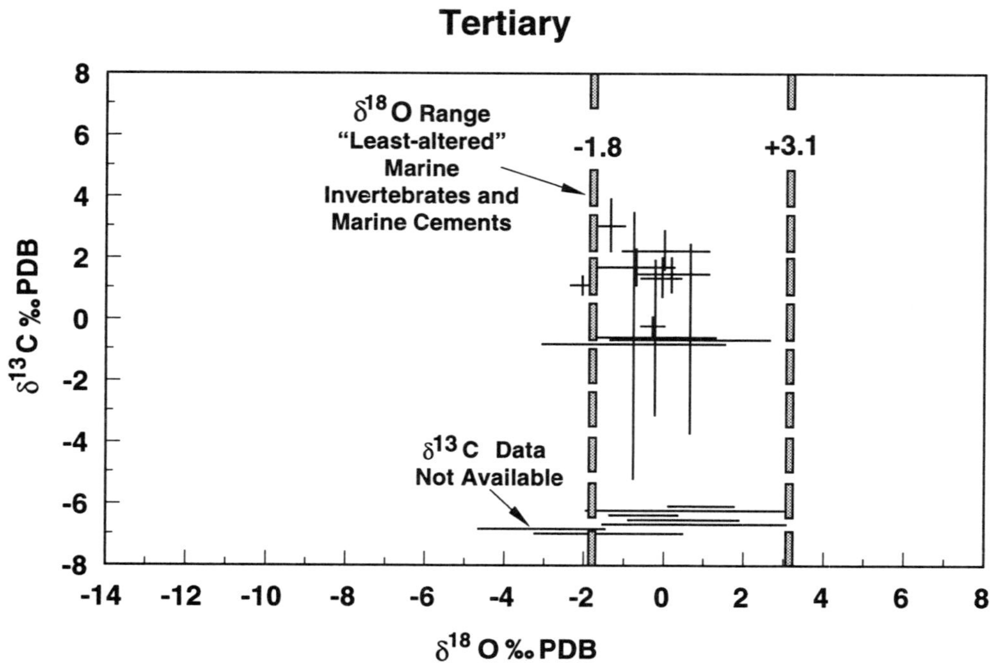

Figure II-3

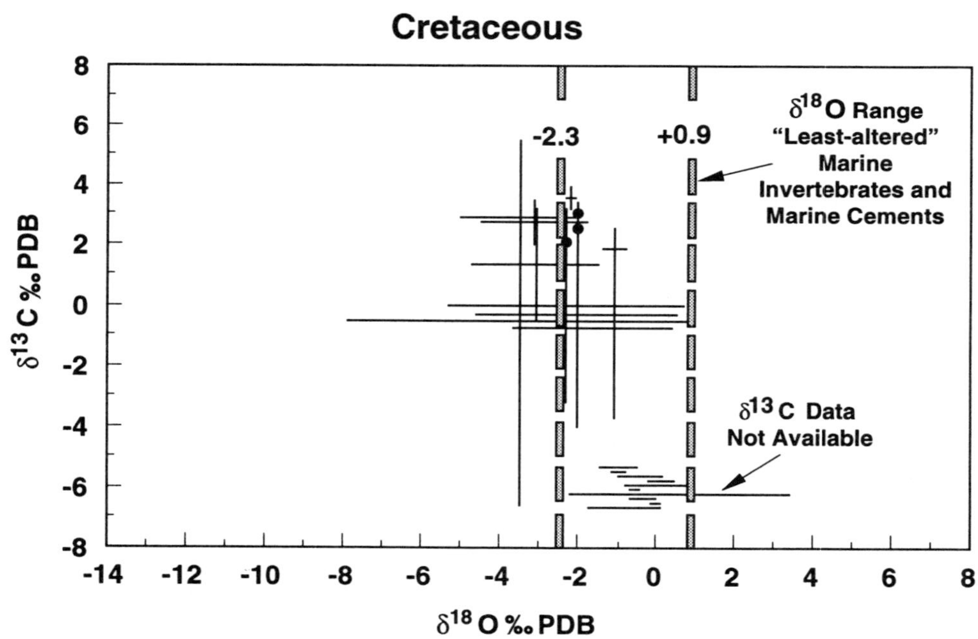

Figure II-4

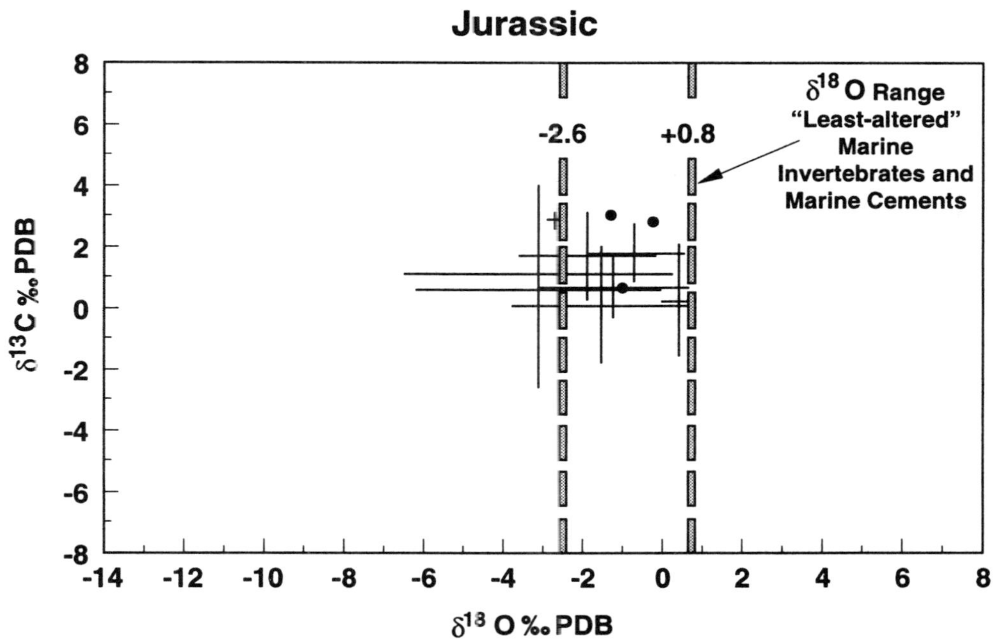

Figure II-5

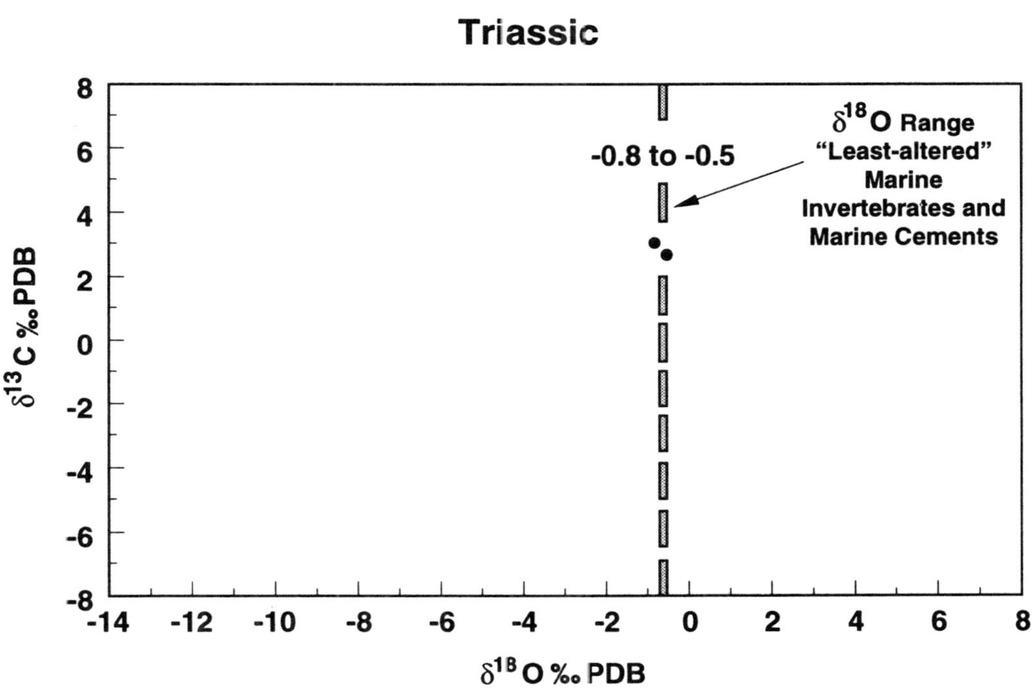

Figure II-6

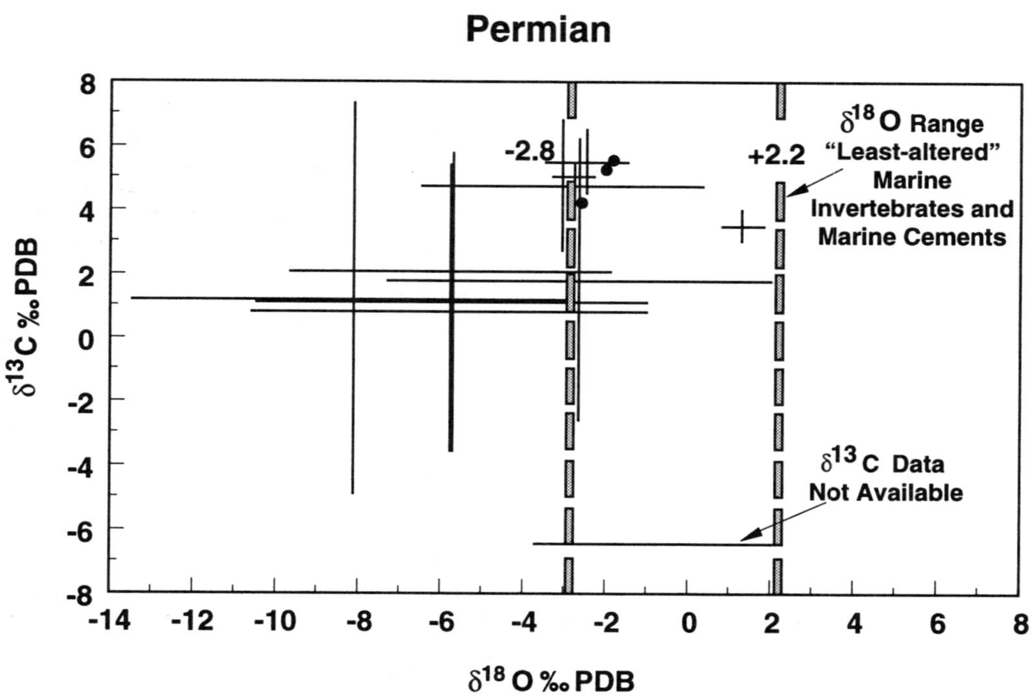

Figure II-7

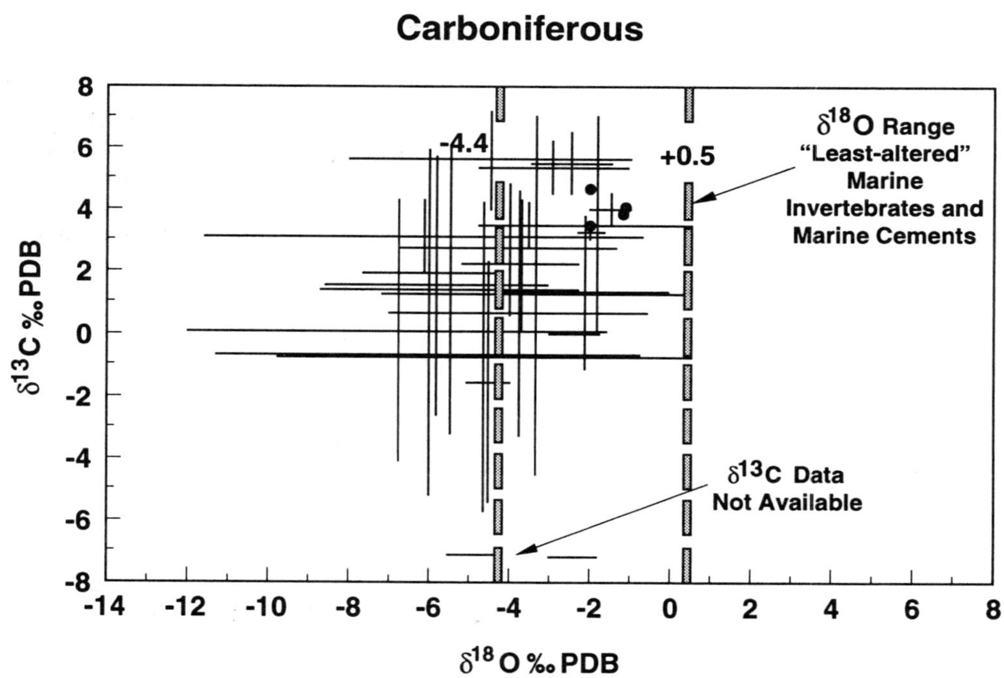

Figure II-8

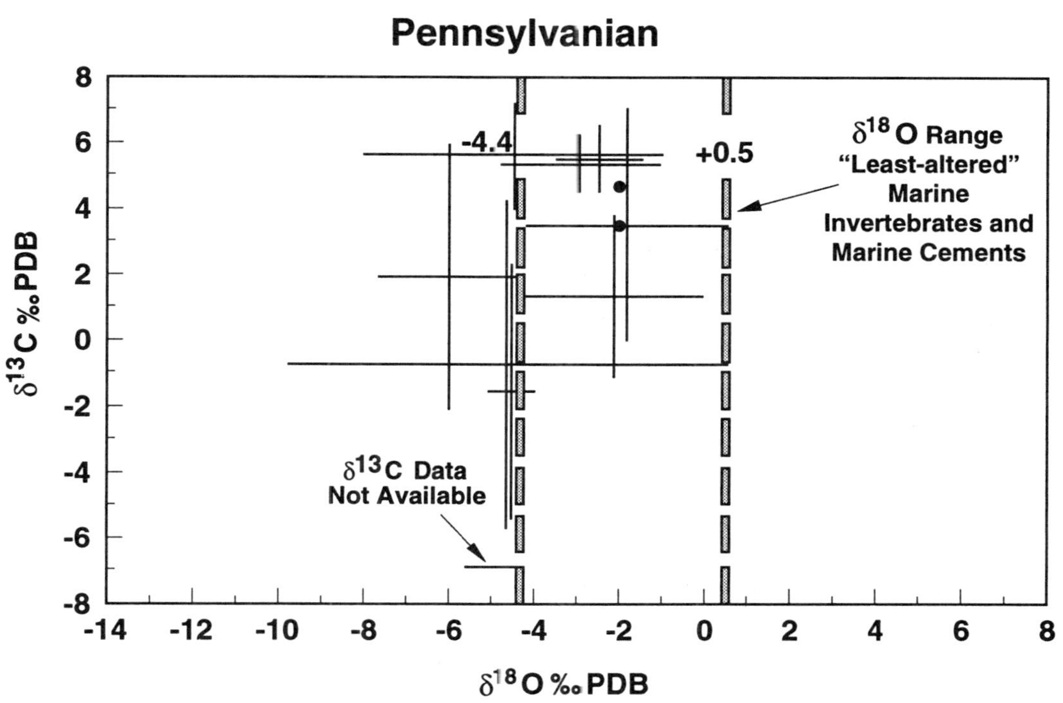

Figure II-9

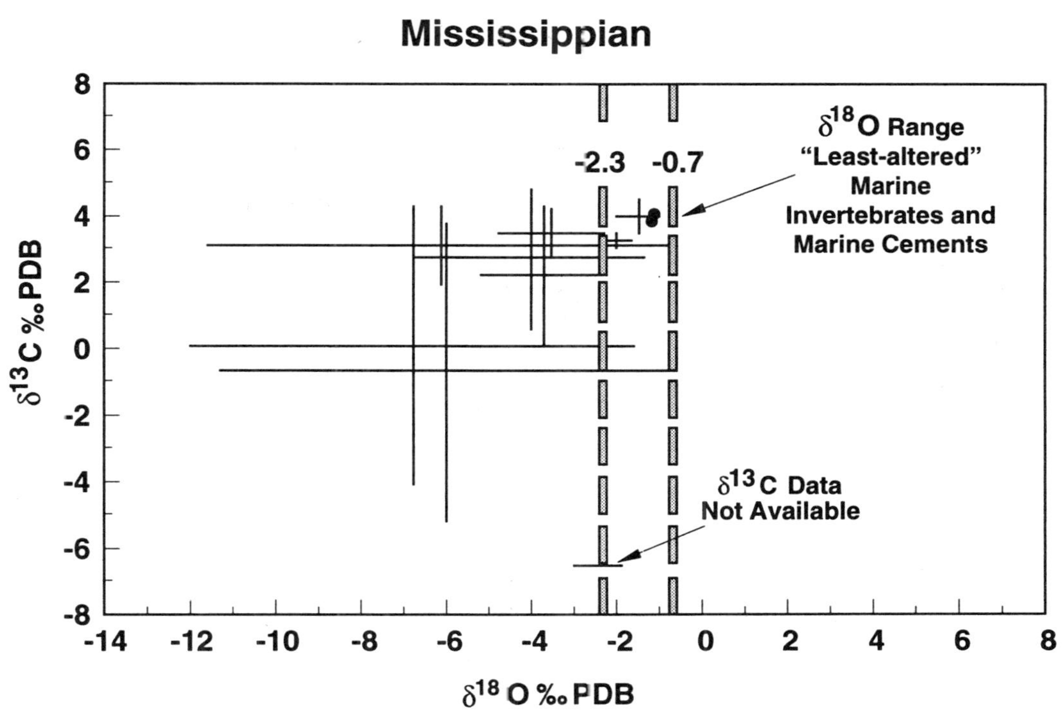

Figure II-10

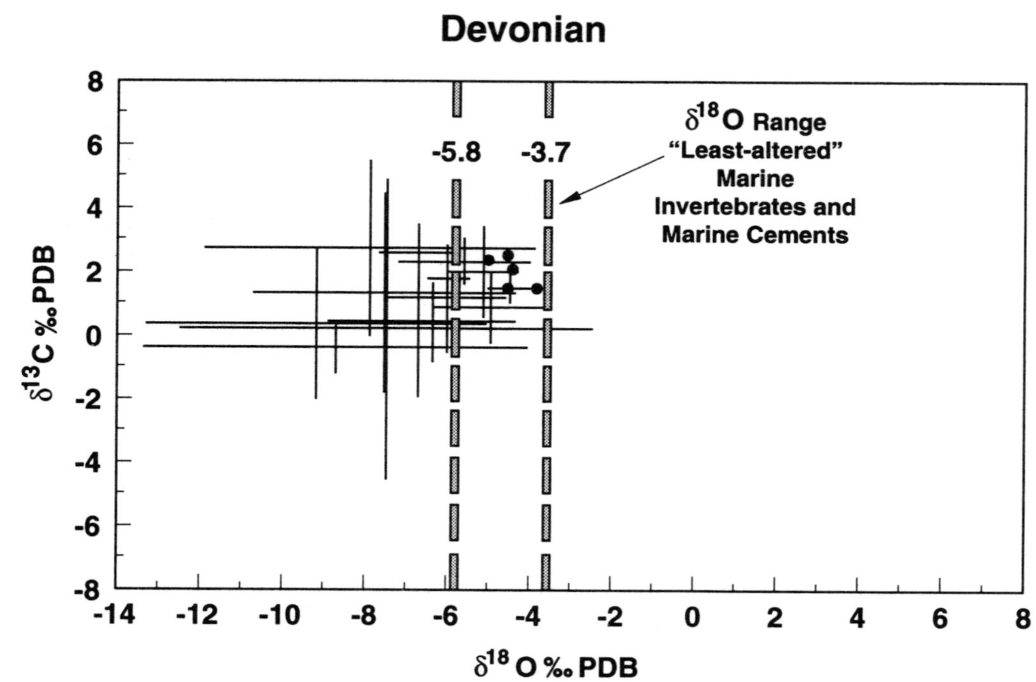

Figure II-11

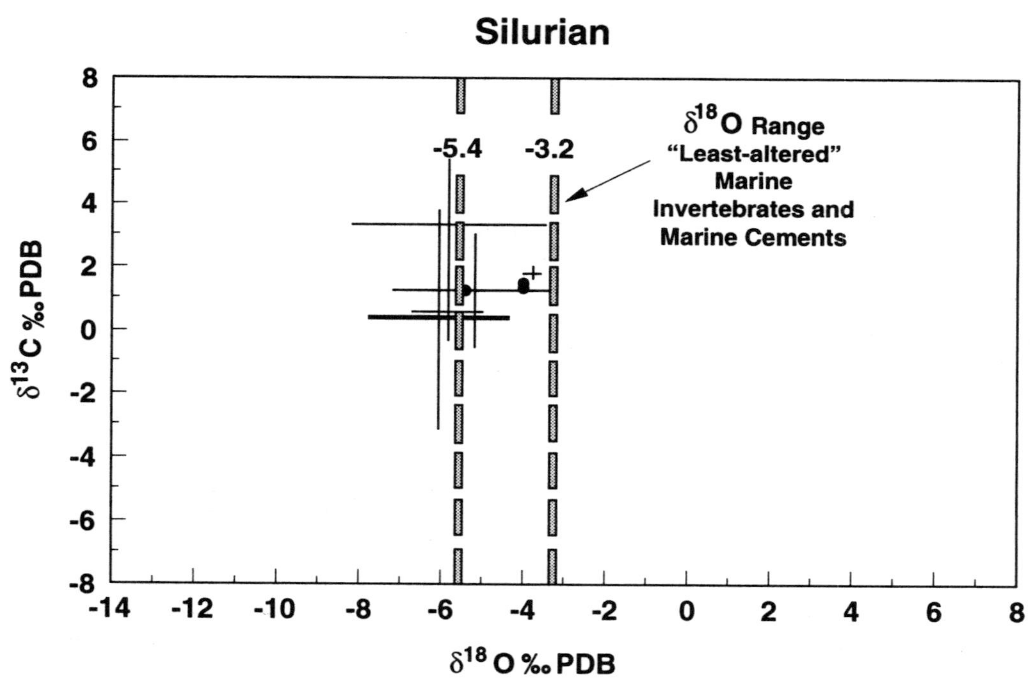

Figure II-12

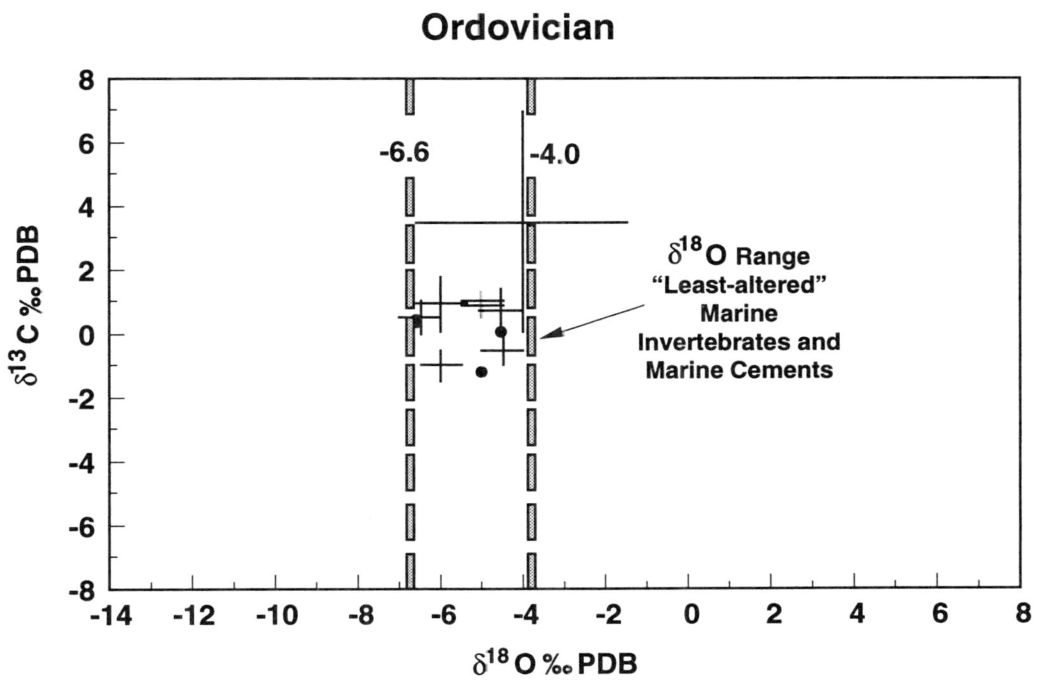

Figure II-13

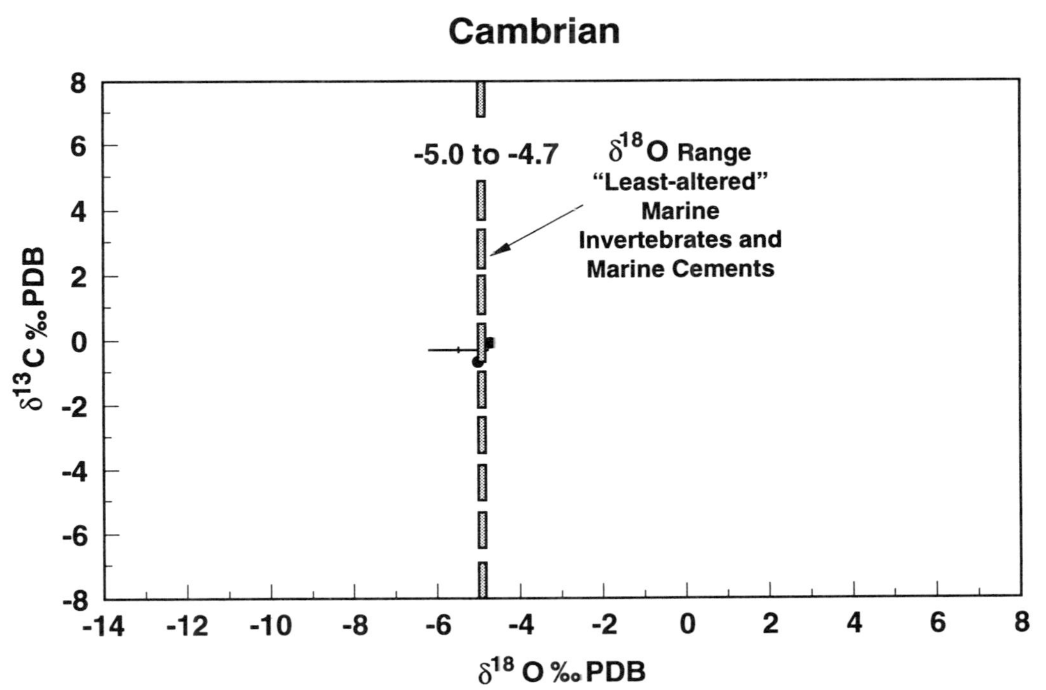

Figure II-14

APPENDIX III. INTERPRETATION OF DOLOMITES BY COMPARISON TO OTHER DOLOMITES

DESCRIPTION OF THE TECHNIQUE

The origin of an unknown dolomite can be determined by comparing its isotopic composition to isotope ranges for low temperature and high temperature dolomites. This technique was developed at Chevron Oil Field Research Co. by J. R. Allan. The construction of the plots used and the rationale behind them are described in this Appendix.

Early, low temperature dolomite is defined as dolomite that formed in supratidal, sabkha, reflux, marine, and marine-meteoric mixing zone environments at or near the earth's surface at low temperature from water that ranged from brackish to hypersaline. Carbon and oxygen isotope ranges for early low temperature dolomite from a variety of environments, localities, and ages are presented in Table III-1 and are summarized in Fig. III-1. Criteria used by the various authors to assign an early origin to the dolomites in Table III-1 include depositional environment of the dolomitized sediment, association with supratidal evaporites, petrography, presence of single phase fluid inclusions, and other geochemical parameters, such as trace elements. All but 4 of the 64 examples of low temperature dolomite have $\delta^{18}O$ values more positive than about -6.5‰ PDB.

Late, high temperature dolomite forms during burial, typically as replacement phases; as late-stage dolomite cement that fills fractures, vugs, and intergranular pore space; and as hydrothermal gangue dolomite associated with ore deposits. Late dolomite commonly contains two-phase fluid inclusions with homogenization temperatures of >50°C. Many have curved or "saddle-shaped" crystal faces and display undulatory extinction—both features are generally associated with precipitation at elevated temperatures. Carbon and oxygen isotope ranges for late high temperature dolomites from a variety of environments, localities, and ages are presented in Table III-2 and are summarized in Fig. III-2. All but 1 of the 54 examples of high temperature dolomite have $\delta^{18}O$ values which are more negative than about -2.5‰ PDB.

From Figs. III-1 and III-2, one can conclude that dolomite which is more positive than -2.5‰ PDB most probably had a low temperature origin and dolomite which is more negative than -6.5‰ PDB most probably had a high temperature origin. When Figs. III-1 and III-2 are overlain on one another, a region of overlap occurs between -2.5 and -6.5‰. In this region, a dolomite could have had either origin. This relationship is displayed graphically in Fig. 20. A low temperature or high temperature origin can be assigned to many unknown dolomites by plotting their $\delta^{18}O$ and $\delta^{13}C$ values on this figure.

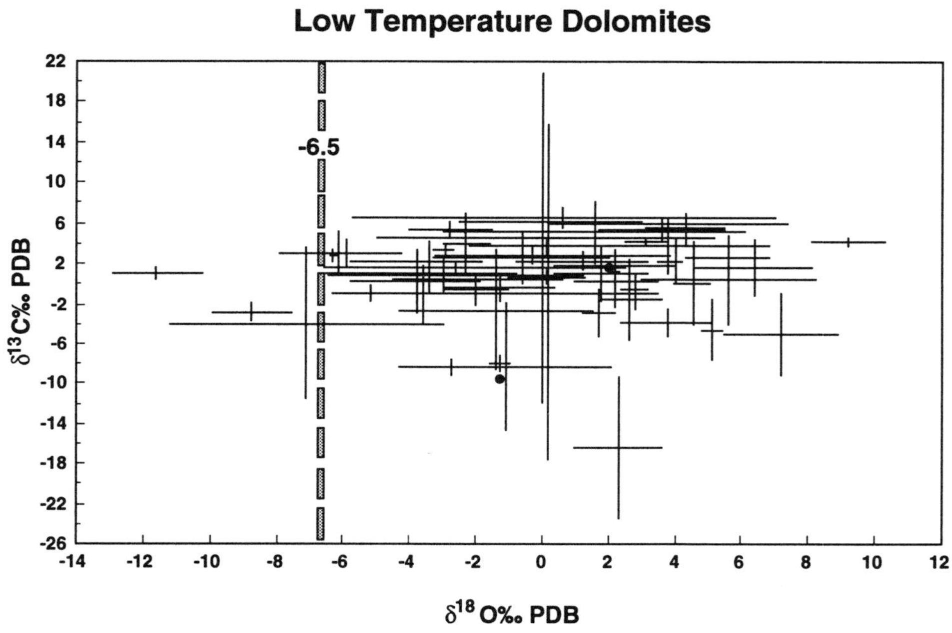

Figure III-1. $\delta^{18}O$ and $\delta^{13}C$ ranges for early, low temperature dolomites which formed in supratidal, sabkha, reflux, hypersaline lagoon, schizohaline ephemeral lake (Coorong), marine, and marine-meteoric mixing zone environments. Several papers reported only average values; these are shown as points. All but 4 of 64 examples of low temperature dolomite shown in figure have $\delta^{18}O$ values more positive than about -6.5 ‰ PDB. This value can be considered the effective lower limit in $\delta^{18}O$ for most low temperature dolomite. Data from which the figure was constructed are compiled in Table III-1.

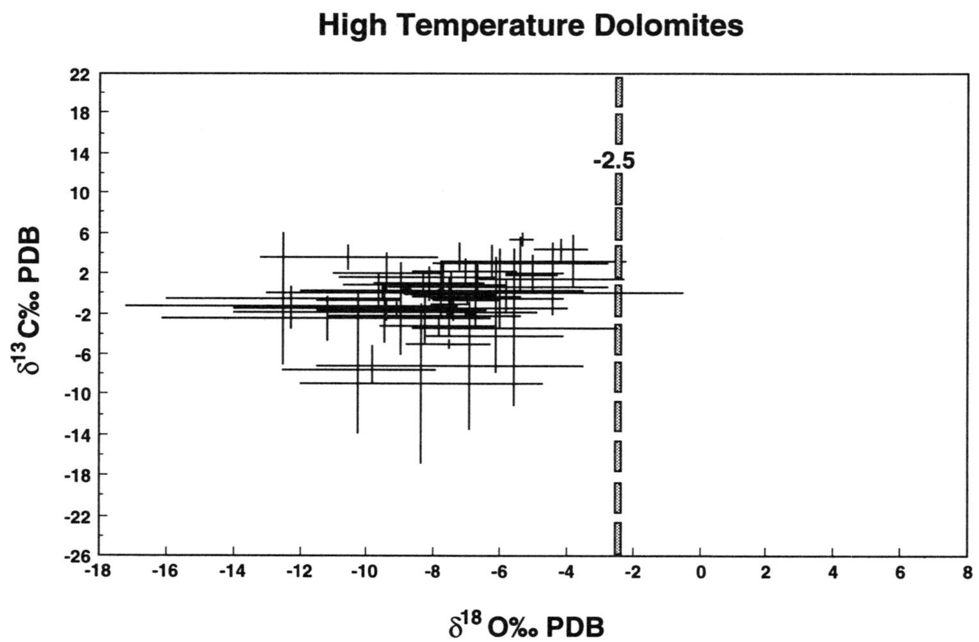

Figure III-2. $\delta^{18}O$ and $\delta^{13}C$ ranges for late, high temperature dolomite which formed as burial replacement phases containing high temperature fluid inclusions; as late-stage dolomite cement which fills fractures, vugs, and intergranular pore space; and as hydrothermal gangue dolomite associated with ore deposits. All but 1 of these 54 examples of high temperature dolomite have $\delta^{18}O$ values more positive than about -2.5 ‰ PDB. This value can be considered the effective upper limit in $\delta^{18}O$ for most high temperature dolomite. Data from which the figure was constructed are compiled in Table III-2.

LIMITATIONS OF THE TECHNIQUE

Several problems present themselves when dealing with the low temperature data set. The evidence used to establish an early dolomitizing environment is sometimes subjective, particularly in the absence of early evaporites. Furthermore, early, low temperature dolomite can alter during burial by crystal ordering, exsolution, and solution-reprecipitation (Reeder, 1981) and this can lead to a change in its isotopic composition (Gao, 1990; Gao and Land, 1991). Although the body of low temperature dolomite data indicates that partial rather than complete recrystallization is the rule (i.e., early dolomite with $\delta^{18}O < -6.5‰$ is rare) complete or nearly complete recrystallization does occur in several instances and can lead to erroneous interpretations. In addition, the $\delta^{18}O$ signature of marine carbonate material was significantly lighter in the early Paleozoic (Cambrian through Devonian) than in the rest of the Phanerozoic (see Appendix II). Early low temperature dolomite that acquired this light $\delta^{18}O$ signature during formation and then partially recrystallized during burial might easily have shifted into the high temperature field. Indeed, the three unusually light examples shown in Fig. III-1 are Cambrian,Ordovician, and Silurian in age. One needs to be aware of this problem when evaluating early Paleozoic dolomites.

In addition, two general sources of uncertainty affect both low and high temperature dolomite data when comparing data from different localities and laboratories:

The first uncertainty is introduced in the measurement of $\delta^{18}O_{dol}$. The fractionation factor (α) for CO_2 liberated from dolomite in phosphoric acid varies as a function of dolomite stoichiometry. Land (1980) showed that $10^3 \ln\alpha$ (the fractionation between oxygen in dolomite and oxygen in CO_2 liberated from dolomite by phosphoric acid reaction, expressed in parts per mil) can vary from 11.51‰ to 12.58‰ for natural dolomites and protodolomites. This variation occurs because dolomite is a solid solution series with a range of chemical compositions (stoichiometries) and variable crystal ordering. Because dolomite crystals can have individual growth zones with very different stoichiometries (e.g., Ward and Halley, 1985), $10^3 \ln\alpha$ variations on the order of 1‰ within a dolomite generation precipitated from the same fluid at the same temperature can be produced by stoichiometry alone.

Second, because α varies as a function of stoichiometry and crystal ordering, it is necessary to choose an arbitrary value for $10^3 \ln\alpha$ when calculating dolomite $\delta^{18}O$ values from mass spectrometer data in the laboratory. Two camps have arisen: those who use a dolomite-CO_2 $10^3 \ln\alpha$ value (11.03‰ for $Ca_{50}Mg_{50}(CO_3)_2$ as reported in Friedman and O'Neil, 1977) and those who use the calcite-CO_2 $10^3 \ln\alpha$ value (10.20‰ as reported in Friedman and O'Neil, 1977). As a result, a single dolomite sample can be assigned two $\delta^{18}O$ values which depart from one another by approximately 0.8‰ depending on the the $10^3 \ln\alpha$ value chosen by a particular laboratory. No attempt was made in this compilation to identify and correct data reported in the literature which were reduced using the calcite-CO_2 $10^3 \ln\alpha$ value. In fact, because many papers report data without identifying the laboratory which ran it, it would be impossible to do so.

The borders to the low temperature and high temperature dolomite fields are based on empirical relationships derived from aggregates of data. Because of the uncertainties involved in determining the original environments of dolomites used to establish the fields, the partial recrystallization of some low temperature dolomites during burial, and the poorly known α for dolomite, the field boundaries should not be taken as absolute, but merely as useful guides for interpreting dolomite origin. Although the general positions of the field boundaries are well-supported by the data, exceptions to the rule should be expected to occur. Therefore carbon and oxygen isotopes should be used in conjunction with other lines of evidence when evaluating dolomite origin. Even with the many uncertainties discussed above, the lower limit for $\delta^{18}O$ of low temperature dolomite (about -6.5‰ PDB) and the upper limit for $\delta^{18}O$ of high temperature dolomite (about -2.5‰ PDB) seem to hold for the great majority of field examples reported in Tables III-1 and III-2. Therefore, this approach represents an improvement over the current practice of plotting isotopic compositions of dolomite on scattergrams, comparing them to a few arbitrarily chosen examples from other studies, or simply describing them as "light" or "heavy". At the same time this method is more prudent than calculating dolomitization temperature from a highly speculative estimation of $\delta^{18}O_{water}$.

Table III-1. Carbon and oxygen isotope ranges for early low temperature dolomite.

Interpreted Environment*	Reference	Age**	Location	$\delta^{18}O$ ‰ PDB	$\delta^{13}C$ ‰ PDB
Marine	Behrens and Land (1972)	Holocene	Baffin Bay, Texas, USA	+4.1 to +5.2	-4.1 to +4.1
Marine, recrystallized in meteoric water	Gao (1990)	Cambrian	Arbuckle Grp., Oklahoma, USA	-12.9 to -10.2	+0.6 to +1.6
Marine	Kier (1973)	Holocene	Baffin Bay, Texas, USA	+4.8 to +5.4	-7.5 to -1.6
Marine	Mitchell et al. (1987)	Holocene	Discovery Bay, Jamaica	+2.0	+1.6
Marine (fabric-mimicking)	Pedone (1990)	Devonian	Canning Basin, Australia	-1.0 to +1.3	-0.1 to +1.3
Marine or marine-meteoric mixing zone (both crystal-line mimetic and micro-sucrosic)	Supko (1977) and Dawans and Swart (1988)	Miocene-Pliocene	San Salvador, W.I.	+0.4 to +4.1	+0.3 to +3.4
Normal marine lagoon (Estheria shales)	Tan and Hudson (1971)	Jurassic	Great Estuarine Series, Scotland	-5.8 to -1.8	+1.4 to +3.2
Deep water marine	Saller (1984)	Eocene	Enewetak Atoll	+2.1 to +2.7	+2.2 to +2.4
Deep water marine	Burns and Baker (1987)	Miocene	Monterey Fm., California, USA	-5.0 to +5.1	-11.8 to +20.8
Deep water marine	Degens and Epstein (1964)	Tertiary	Gulf of Mexico (Mohole)	+2.1 to +2.3	+1.1 to +1.4
Deep water marine	Graber and Lohmann (1989)	Pennsylvanian-Permian	Horquilla Fm., New Mexico, USA	-6.5 to +0.5	-0.8 to +2.6
Deep water marine	Mullins et al. (1988)	Miocene, Pliocene, Pleistocene	DSDP & ODP Sites, offshore Florida, USA	+4.3 to +6.8	+1.6 to +3.5
Deep water marine	Pisciotto and Mahoney (1981)	Miocene-Pliocene	DSDP Leg 63, offshore Mexico and California, USA	-3.0 to +3.5	-17.6 to +15.8
Marine-meteoric mixing zone	Badiozamani (1973)	Ordovician	Mifflin Mbr., Platteville Fm., Wisconsin, USA	-6.3 to -4.0	-1.8 to -0.2
Marine-meteoric mixing zone (Dolomite I)	Banner et al. (1988)	Mississippian	Burlington – Keokuk Fm., Illinois and Missouri, USA	-2.2 to +2.5	-0.9 to +4.0
Marine-meteoric mixing zone recrystallized in subsurface fluids (Dolomite II)	Banner et al. (1988)	Mississippian	Burlington – Keokuk Fm., Illinois & Missouri, USA	-6.6 to -0.2	-1.0 to +4.1

* Where more than one type of dolomite is reported in paper, author's description of the particular dolomite listed in this Table is shown in parentheses.

** Age of host rock. Dolomite may be younger.

Table III-1 - 2

Interpreted Environment*	Reference	Age**	Location	$\delta^{18}O$ ‰ PDB	$\delta^{13}C$ ‰ PDB
Marine-meteoric mixing zone (first generation replacive dolomites)	Cercone and Lohmann (1985)	Silurian	Niagara Fm., Michigan, USA	-8.0 to -4.3	+0.9 to +5.2
Marine-meteoric mixing zone	Choquette and Steinen (1980)	Mississippian	Ste. Genevieve Fm.,Illinois, USA	-0.8 to +3.2	+1.4 to +3.1
Marine-meteoric mixing zone	Dunham and Olson (1980)	Ordovician-Silurian	Hanson Creek Fm., Nevada, USA	-6.7 to -0.8	0.0 to +2.2
Marine-meteoric mixing, zone, reflux, or deep water marine	Enos (1988)	Cretaceous	Poza Rica trend, Mexico	-3.0 to -1.6	+1.1 to +7.0
Marine-meteoric mixing zone	Fritz (1967)	Jurassic	Malm, Germany	-5.8 to -1.9	-2.9 to +3.3
Marine-meteoric mixing zone	Holail et al. (1988)	Cretaceous	El Heiz and El Hefhuf Fms., Egypt	-3.3 to +2.0	0.0 to +5.2
Marine-meteroic mixing zone	Humphrey (1988)	Pleistocene	Golden Grove terrace, Barbados, W.I.	+1.0 to +3.6	-23.5 to -9.4
Marine-meteoric mixing zone?	Land and Epstein (1970)	Pleistocene	Falmouth Fm., Jamaica, W.I.	-3.1 to -2.4	-9.1 to -7.5
Marine or marine-meteoric mixing zone	Land (1973a)	Pleistocene	Falmouth Fm., Jamaica, W.I.	-1.6 to -0.9	-8.7 to -7.1
Marine or marine-meteoric mixing zone?	Land (1973b)	Pliocene	Aruba, Bonaire, & Curacao, Netherlands Antilles	+1.7 to +3.6	-5.7 to +2.4
Marine or marine-meteoric mixing zone	Land (1973b)	Pleistocene	Hope Gate Fm., Jamaica, W.I.	+0.4 to +3.2	-1.8 to +3.8
Marine-meteoric mixing zone	Land et al. (1975)	Eocene	El Naqb Fm., Egypt	-3.0 to +0.4	-1.8 to +1.1
Marine-meteoric mixing zone	Land et al. (1975)	Ordovician	Allen Bay Fm., NWT, Canada	-4.5 to +0.6	+0.1 to +0.9
Marine-meteoric mixing zone	Magaritz et al. (1980)	Quaternary	Coastal Plain, Israel	-1.3	-9.6
Marine-meteoric mixing zone (coastal oolitic barrier)	Peryt and Magaritz (1990)	Permian	Zechstein Fm., Poland	-5.7 to +7.0	+5.5 to +7.6
Marine-meteoric mixing zone	Pierson (1982)	Tertiary	Southeastern Bahamas, W.I.	+1.0 to +3.4	-2.2 to +2.7
Marine-meteoric mixing zone	Sibley (1980)	Pliocene	Seroe Domi Fm., Bonaire, N. A.	+2.0 to +4.1	-
Marine-meteoric mixing zone	Ward and Halley (1985)	Pleistocene	Yucatan, Mexico	+1.1 to +2.4	+0.5 to +3.2

Table III-1 - 3

Interpreted Environment*	Reference	Age**	Location	$\delta^{18}O$ ‰ PDB	$\delta^{13}C$ ‰ PDB
Coorong schizohaline ephemeral lake	Clayton et al. (1968)	Holocene	Coorong region, Australia	+3.9 to +4.2	+0.1 to +4.5
Coorong schizohaline emphemeral lake	Rosen et al. (1988)	Holocene	Pellet Lake, Australia	+4.6 to +8.1	-1.1 to +4.3
Coorong schizohaline ephemeral lake	Warren (1988)	Holocene	Salt Creek Region, Australia	+3.0 to +8.2	-4.0 to +4.8
Sabkha	Kinsman and Patterson (1973)	Holocene	Abu Dhabi	+2.5 to +3.7	+4.0 to +4.4
Sabkha (dolomite associated with evaporites)	M'Rabet (1981)	Cretaceous	Bouhedma Fm., Tunisia	-1.1 to +1.2	-0.4 to +2.0
Sabkha	McKenzie (1981)	Holocene	Abu Dhabi	-0.4 to +3.9	+2.0 to +3.7
Sabkha and marine hypersaline lagoon	Moore et al. (1988)	Jurassic	Buckner Fm., Texas, USA	-2.2 to +2.4	+2.9 to +4.8
Sabkha	Peryt and Magaritz (1990)	Permian	Zechstein Fm., Poland	+0.8 to +6.8	+1.1 to +6.5
Sabkha	Pierre et al. (1984)	Holocene	Ojo de Liebre, Mexico	+2.4 to +5.1	-5.3 to -2.4
Sabkha, recrystallized in meteoric water	Gao and Land (1991)	Ordovician	Arbuckle Grp., Oklahoma, USA	-9.9 to -7.6	-3.6 to -1.9
Sabkha overprinted by mixing zone	Randazzo and Cook (1987)	Eocene	Avon Park Fm., Florida, USA	+3.5 to +4.2	+1.2 to +3.1
Marine hypersaline pond	Aharon et al. (1976)	Holocene	Gulf of Elat, Israel	+8.1 to +10.3	+3.8 to +4.5
Evaporitic marine	Bellanca and Neri (1986)	Miocene	Messinian, central Sicily	+5.5 to +8.9	-9.1 to -0.9
Marine evaporitic lagoon	Peryt and Magaritz (1990)	Permian	Zechstein Fm., Poland	+0.2 to +7.4	+4.9 to +6.9
Hypersaline marine lagoon (Ostracod Limestone)	Tan and Hudson (1971)	Jurassic	Great Estuarine Series, Scotland	-4.3 to +2.1	-14.7 to -1.9
Hypersaline lagoon (early diagenetic dolomite)	Veizer (1970); Veizer and Compston (1974); Veizer and Hoefs (1976)	Triassic	Western Carpathians, Czechoslovakia	-4.6 to -2.6	-3.9 to +1.8
Supratidal	Carballo et al. (1987)	Holocene	Sugarloaf Key, Florida, USA	+2.4 to +3.2	-2.4 to +1.1
Intertidal-supratidal (marine)	Degens and Epstein (1964)	Triassic	Karn, Nor & Rhat, Austria	-3.2 to +2.6	+2.1 to +3.6
Intertidal-supratidal (marine and caliche)	Degens and Epstein (1964)	Triassic	Muschelkalk & Buntsandstein, Germany	-4.3 to +1.5	-8.6 to +3.4
Supratidal (Group I)	Fritz and Jackson (1972)	Devonian	Pine Point, NWT, Canada	-3.0 to -1.0	-2.0 to +1.0
Supratidal and intertidal	Leary and Vogt (1986)	Permian	San Andres Fm., Texas, USA	+3.1 to +5.5	+4.0 to +7.0
Supratidal	Mazzullo et al. (1987)	Holocene	Ambergris Cay, Belize	+1.2 to +2.2	-5.2 to -0.6

Table III-1 - 4

Interpreted Environment*	Reference	Age**	Location	$\delta^{18}O$ ‰ PDB	$\delta^{13}C$ ‰ PDB
Supratidal or reflux	O'Shea et al. (1988)	Silurian	Salina Fm., Michigan, USA and Canada	-11.2 to -3.0	-11.5 to +3.5
Supratidal or marine	Qing and Mountjoy (1989)	Devonian	Keg River Fm., Alberta, Canada	-6.4 to -6.2	+2.3 to +3.3
Supratidal or marine (fine crystalline)	Anderson (1985)	Devonian	Nisku Fm., Alberta, Canada	-3.3 to -2.6	+2.9 to +3.9
Reflux (second generation dolomites)	Cercone and Lohmann (1985)	Silurian	Niagara Fm., Michigan, USA	-6.8 to -5.0	+1.7 to +4.3
Reflux	Garber et al. (1990)	Permian	Tansill & Yates Fms., New Mexico, USA	-2.5 to +3.0	+4.5 to +8.0
Reflux (subtidal)	Leary and Vogt (1986)	Permian	San Andres Fm., Texas, USA	+1.7 to +5.5	+4.3 to +6.3
Reflux	Moore et al. (1988)	Jurassic	Smackover Fm., Texas, USA	-4.0 to -1.5	+4.6 to +6.2
Reflux (deeper ramp)	Peryt and Magaritz (1990)	Permian	Zechstein Fm., Poland	-3.0 to +6.1	+2.1 to +8.1

Table III-2 — 1

Table III-2. Carbon and oxygen isotope ranges for late high temperature dolomites.

Interpreted Environment	Author's Description	Reference	Age*	Location	$\delta^{18}O$ ‰ PDB	$\delta^{13}C$ ‰ PDB	Evidence for High Temperature Origin
Burial	Baroque dolomite cement	Alsharhan & Williams (1987)	Cretaceous	Shuaiba Fm., Abu Dhabi	-9.7 to -4.8	+2.4 to +4.9	Saddle dolomite texture.
Burial	Matrix selective & pervasive dolomite	Anderson (1985)	Devonian	Nisku Fm., Alberta, Canada	-5.4 to -2.2	+0.6 to +5.8	Increases in abundance down structural dip (i.e., postdates structural movement).
Burial	Baroque dolomite	Anderson (1985)	Devonian	Nisku Fm., Alberta, Canada	-5.8 to -4.3	+0.1 to +3.7	Fluid inclusion temperatures.
Burial	Ankerite cement	Boles (1978)	Eocene	Wilcox Fm., Texas, USA	-11.5 to -8.2	-9.0 to -5.2	Found only in more deeply buried samples (T>125°C).
Burial	Host dolomite	Broomhall & Allan (1987)	Jurassic & Cretaceous	Eastern Saudi Arabia	-9.3 to -2.7	-3.5 to +4.4	Host & baroque dolomites in each sample have nearly identical $\delta^{18}O$.
Burial	Baroque dolomite cement	Broomhall & Allan (1987)	Jurassic & Cretaceous	Eastern Saudi Arabia	-9.4 to -4.1	-2.2 to +2.9	Fluid inclusion temperatures.
Burial	Vein-filling dolomite	Budai et al. (1987)	Mississippian	Madison Group, Wyoming/Utah, USA	-12 to -3.5	-2.5 to +3	Late in paragenetic sequence.
Burial	Fracture-filling dolomite	Carpenter et al. (in prop.)	Ordovician	Trenton Fm., Michigan, USA	-11.5 to -7.4	-1.5 to +1.8	Fluid inclusion temperatures.
Burial	Fracture-filling dolomite	Cercone & Lohmann (1987)	Silurian	Niagara Fm., Michigan, USA	-5.7 to -5.0	+4.9 to +5.9	Fluid inclusion temperatures; petroleum inclusions.
Burial	Ferroan dolomite cement	Choquette (1971)	Mississippian	Ste. Genevieve Fm., Illinois, USA	-7.3 to -5.0	-3.3 to +3.5	Saddle dolomite texture; late in paragenetic sequence.
Burial	Replacement dolomite	Kaufman (1989)	Devonian	Swan Hills Fm., Alberta, Canada	-8.6 to -5.5	+1.2 to +3.4	Postdates early stylolites & burial mosaic calcites.
Burial	Saddle dolomite	Kaufman (1989)	Devonian	Swan Hills Fm., Alberta, Canada	-8.7 to -6.2	-2.5 to +2.3	Fluid inclusion temperatures.

* Age of host rock. Dolomite may be younger.

Table III-2 — 2

Interpreted Environment	Author's Description	Reference	Age*	Location	$\delta^{18}O$ ‰ PDB	$\delta^{13}C$ ‰ PDB	Evidence for High Temperature Origin
Burial	Ankerite cement	Land et al. (1987)	Eocene	Frio Fm., Texas, USA	-10.5 to -3.5	-13.5 to -1	Late in paragenetic sequence.
Burial	Ankerite cement	Land & Dutton (1978)	Pennsylvanian	Strawn Fm., Texas, USA	-8.1 to -6.9	-3.8 to +1.4	Late in paragenetic sequence.
Burial	Ankerite cement	Land & Fisher (1987)	Eocene	Wilcox Fm., Texas, USA	-12.5 to -8	-14 to -1	Replaces calcite, only in more deeply buried samples.
Burial	Matrix dolomite	Machel (1985)	Devonian	Nisku Fm., Alberta, Canada	-6.6 to -2.3	-2.1 to +5.0	Postdates stylolites; stratigraphic relations suggest upwardly moving fluids.
Burial	Saddle dolomite	Machel (1985)	Devonian	Nisku Fm., Alberta, Canada	-8.6 to -2.5	-11.2 to +4.3	Fluid inclusion temperatures.
Burial	Mosaic dolomite	Mattes & Mountjoy (1980)	Devonian	Cairn & Southesk Fms., Alberta, Canada	-8.0 to -2.8	+0.2 to +5.6	Stratigraphic relations preclude early dolomitization.
Burial	White sparry dolomite	Mattes & Mountjoy (1980)	Devonian	Cairn & Southesk Fms., Alberta, Canada	-11.5 to -9.0	-1.5 to +0.2	Stratigraphic relations preclude early dolomitization.
Burial	Post-compaction dolomite	Moore et al. (1988)	Jurassic	Smackover Fm., Arkansas & Texas, USA	-5.0 to -3.4	+3.4 to +5.3	Late in paragenetic sequence.
Burial	Saddle dolomite	Moore et al. (1988)	Jurassic	Smackover Fm., Arkansas & Texas, USA	-7.8 to -4.7	+1.6 to +4.8	Saddle dolomite texture.
Burial	Rhombs, patches, & matrix dolomites	Qing & Mountjoy (1989)	Devonian	Keg River Fm., Alberta, Canada	-7.2 to -6.1	+0.3 to +2.8	Postdates some stylolites & burial calcite cements.
Burial	Saddle dolomite	Qing & Mountjoy (1989)	Devonian	Keg River Fm., Alberta, Canada	-10.8 to -8.0	+1.5 to +2.3	Late in paragenetic sequence; fills late fractures.
Burial	Ferroan dolomite	Sternback & Friedman (1986)	Ordovician-Devonian	Hunton Grp., Oklahoma, USA	-6 to -4.5	—	Percent dolomite increases with depth in same horizon.
Burial	Fracture-filling dolomite	Viau & Oldershaw (1984)	Devonian	Swan Hills Fm., Alberta, Canada	-11.0 to -7.8	+0.2 to +3.9	Saddle dolomite texture; late in paragenetic sequence.

Table III-2 — 3

Interpreted Environment	Author's Description	Reference	Age*	Location	$\delta^{18}O$ ‰ PDB	$\delta^{13}C$ ‰ PDB	Evidence for High Temperature Origin
Burial	Coarsely crystalline dolomite	Zenger (1983)	Devonian	Lost Burro Fm., California, USA	-8.1 to -4.1	-1.2 to +0.2	Stratigraphic & petrographic relationships.
Burial	Replacement dolomite	Zenger & Dunham (1988)	Silurian-Devonian	New Mexico, USA	-9.1 to -6.3	0 to +3.2	Fluid inclusion temperatures.
Burial	Saddle dolomite	Zenger & Dunham (1988)	Silurian-Devonian	New Mexico, USA	-7.8	0	Fluid inclusion temperatures.
Deep burial	Pink & white saddle dolomite	Friedman (1989)	Silurian	Lockport Fm., New York, USA	-11.1 to -9.6	—	Fluid inclusion temperatures.
Deep burial	Coarse crystalline dolomite (Type 3)	Lee & Friedman (1987)	Ordovician	Ellenberger Fm., Texas, USA	-8.2 to -4.1	-7.7 to -1.0	Undulatory extinction; replaces across stylolites.
Deep burial	Saddle dolomite (Type 4)	Lee & Friedman (1987)	Ordovician	Ellenberger Fm., Texas, USA	-12.0 to -4.7	-16.8 to -1.2	Fluid inclusion temperatures; saddle dolomite texture.
Deep burial	Saddle dolomite	Sellwood et al. (1989)	Jurassic	Great Oolite, England	-13.2 to -7.9	+2.4 to +4.8	Fluid inclusion temperatures.
Deep burial	Baroque dolomite	Stepanek (1987)	Ordovician	El Paso Grp., Texas, USA	-16.1 to -6.3	-4.7 to -0.3	Two-phase fluid inclusions; late in paragenetic sequence; saddle dolomite texture.
Subeconomic MVT mineralization	Saddle dolomite	Buchbinder et al. (1984)	Jurassic	Israel	-10.7 to -5.9	-0.4 to +1.9	Associated with ore minerals.
MVT ore deposit	Reef dolomite	Fritz (1969)	Devonian	Pine Point & Presqu'ile Fms., NWT, Canada	-9.8 to -6.5	-0.7 to +2.6	Host to hydrothermal mineralization.
MVT ore deposit	White vein dolomite	Fritz (1969)	Devonian	Pine Point & Presqu'ile Fms., NWT, Canada	-10.9 to -8.2	-0.3 to +0.8	Associated with ore minerals.
MVT ore deposit	Coarsely crystalline hydrothermal dolomite	Fritz & Jackson (1972)	Devonian	Presqu'ile Fm., NWT, Canada	-11.2 to -8.1	-1.4 to +1.9	Associated with ore minerals.
Possible MVT ore deposit	Replacement dolomite	Gregg (1985)	Cambrian	Bonneterre Fm., Missouri, USA	-8.6 to -6.2	-0.7 to +0.2	Undulatory extinction.

Table III-2 — 4

Interpreted Environment	Author's Description	Reference	Age*	Location	$\delta^{18}O$ ‰ PDB	$\delta^{13}C$ ‰ PDB	Evidence for High Temperature Origin
Possible MVT ore deposit	Dolomite cement	Gregg (1985)	Cambrian	Bonneterre Fm., Missouri, USA	-8.0 to -6.0	-1.2 to -0.2	Undulatory extinction; CL banding correlative with MVT gangue.
MVT ore deposit	Recrystalline dolomite blocks	Haynes (1986)	Ordovician	Mascot-Jefferson City Trend, Tennessee, USA	-7.9 to -7.3	-2.0 to -1.1	Fluid inclusion temperatures; associated with ore minerals.
MVT ore deposit	Coarse rock-matrix breccia	Haynes (1986)	Ordovician	Mascot-Jefferson City Trend, Tennessee, USA	-8.4 to -6.4	-2.6 to -0.8	Fluid inclusion temperatures; associated with ore minerals.
MVT ore deposit	Dolomite spar	Haynes (1986)	Ordovician	Mascot-Jefferson City Trend, Tennessee, USA	-11.5 to -6.7	-2.3 to -1.0	Fluid inclusion temperatures; associated with ore minerals.
MVT ore deposit	Pink sparry dolomite	Long et al. (1986)	Ordovician	Everton Fm., Arkansas, USA	-9.6 to -6.1	-4.3 to -2.2	Fluid inclusion temperatures; associated with ore minerals.
Ore deposit	Microcrystalline dolomite	Beaty & Landis (1990)	Mississippian	Leadville Fm., Colorado, USA	-6.2 to -5.4	-1.7 to +1.2	Wallrock alteration around ore deposit.
Ore deposit	Sucrosic dolomite	Beaty & Landis (1990)	Mississippian	Leadville Fm., Colorado, USA	-11.1 to -5.4	-4.9 to +0.3	Wallrock alteration around ore deposit.
Ore deposit	Coarsely crystalline dolomite	Beaty & Landis (1990)	Mississippian	Leadville Fm., Colorado, USA	-6.9	-2.1	Wallrock alteration around ore deposit.
Ore deposit	Zebra spar	Beaty & Landis (1990)	Mississippian	Leadville Fm., Colorado, USA	-17.2 to -7.3	-3.4 to +0.7	Wallrock alteration around ore deposit.
Ore deposit	Coarsely crystalline dolomite	Engel et al. (1958)	Mississippian	Leadville Fm., Colorado, USA	-8.8	+0.3	Associated with ore minerals.
Ore deposit	Dolomite breccia	Engel et al. (1958)	Mississippian	Leadville Fm., Colorado, USA	-8.8 to -6.3	-5.4 to -4.7	Associated with ore minerals.
Ore deposit	Hydrothermal dolomite	Engel et al. (1958)	Mississippian	Leadville Fm., Colorado, USA	-14.0 to -4.9	-4.9 to +1.0	Associated with ore minerals.
Ore deposit	Coarse pink dolomite	Hall & Friedman (1969)	Paleozoic	Webber Mine, Kansas, USA	-10.4 to -8.4	-2.6 to -0.7	Wallrock alteration around ore deposit.

Table III-2 — 5

Interpreted Environment	Author's Description	Reference	Age*	Location	$\delta^{18}O$ ‰ PDB	$\delta^{13}C$ ‰ PDB	Evidence for High Temperature Origin
Hot hydrothermal fluids during early burial	Host dolomite	Aulstead et al. (1988)	Devonian	Presqu'ile, Manetoe, & Keg River Fms., Canada	-14 to -4	-6 to +3	Similarity in $\delta^{18}O$ and field relationships between host dolomite & dolomite cement.
Hot hydrothermal fluids during early burial	White sparry dolomite cement	Aulstead et al. (1988)	Devonian	Presqu'ile, Manetoe, & Keg River Fms., Canada	-16 to -9	-7 to +6	Fluid inclusion temperatures.
Hydrothermal	Bulk dolomite samples	Cervato (1990)	Jurassic & Cretaceous	Southern Alps, Italy	-13 to -0.5	-3 to +3	Association with volcanic system; dolostones contain high temperature clays.